DATE DUE

2-4-93 Ill	
MAY 1 0 2001	
MAR 1 7 2011	

BIOPHYSICAL AND BIOCHEMICAL ASPECTS OF FLUORESCENCE SPECTROSCOPY

BIOPHYSICAL AND BIOCHEMICAL ASPECTS OF FLUORESCENCE SPECTROSCOPY

Edited by

T. Gregory Dewey

University of Denver
Denver, Colorado

PLENUM PRESS • NEW YORK AND LONDON

Biophysical and biochemical aspects of fluorescence spectroscopy /
  edited by T. Gregory Dewey.
      p.   cm.
  Includes bibliographical references and index.
  ISBN 0-306-43627-2
  1. Fluorescence spectroscopy. 2. Biochemistry--Methodology.
3. Molecular biology--Methodology.   I. Dewey, Thomas Gregory, 1952-
  .
  QP519.9.F56B56 1991
  574.19'285--dc20   90-25341
                                                  CIP

ISBN 0-306-43627-2

© 1991 Plenum Press, New York
A Division of Plenum Publishing Corporation
233 Spring Street, New York, N.Y. 10013

Printed in the United States of America

Contributors

Barbara Baird, Department of Chemistry, Cornell University, Ithaca, New York 14853-6401

John Brumbaugh, School of Biological Sciences, University of Nebraska, Lincoln, Nebraska 68588

Richard A. Cardullo, Worcester Foundation for Experimental Biology, Shrewsbury, Massachusetts 01545

Richard A. Cerione, Department of Pharmacology, Cornell University, Ithaca, New York 14853-6401

T. Gregory Dewey, Department of Chemistry, University of Denver, Denver, Colorado 80208

Maurice R. Eftink, Department of Chemistry, University of Mississippi, University, Mississippi 38677

Jon Erickson, Pierre A. Fish Laboratory, Department of Pharmacology, New York State College of Veterinary Medicine, Cornell University, New York, New York 14853-6401

Paramjit K. Gharyal, Department of Biochemistry, Michigan State University, East Lansing, Michigan 48824

Byron Goldstein, Theoretical Division, Los Alamos National Laboratory, Los Alamos, New Mexico 87544

Dan Grone, Li-Cor, Inc., Lincoln, Nebraska 68504

Theodore L. Hazlett, Department of Biochemistry and Biophysics, John A. Burns School of Medicine, University of Hawaii at Manoa, Honolulu, Hawaii 96822

David Holowka, Department of Chemistry, Cornell University, Ithaca, New York 14853

David M. Jameson, Department of Biochemistry and Biophysics, John A.

vi

Contributors

Burns School of Medicine, University of Hawaii at Manoa, Honolulu, Hawaii 96822

Lian-Wei Jiang, Department of Biochemistry, Michigan State University, East Lansing, Michigan 48824

Lyle Middendorf, Li-Cor, Inc., Lincoln, Nebraska 68504

Robert M. Mungovan, Worcester Foundation for Experimental Biology, Shrewsbury, Massachusetts 01545

Matthew Petersheim, Department of Chemistry, Seton Hall University, South Orange, New Jersey 07079

William J. Phillips, Department of Pharmacology, Cornell University, Ithaca, New York 14853-6401

Richard Posner, Department of Chemistry, Cornell University, Ithaca, New York 14853-6401

Jerry Ruth, Molecular Biosystems, Inc., San Diego, California 92121

Melvin Schindler, Department of Biochemistry, Michigan State University, East Lansing, Michigan 48824

David E. Wolf, Worcester Foundation for Experimental Biology, Shrewsbury, Massachusetts 01545

Preface

Fluorescence spectroscopy has traditionally found wide application in biochemistry and cell biology. Since there are relatively few naturally occurring fluorescent biomolecules, fluorescence spectroscopy offers a combination of great specificity and sensitivity. Historically, these features have been exploited with great success utilizing both intrinsic and extrinsic probes. Recent applications have built upon these traditional strengths and have resulted in the development of new instrumental techniques, novel and convenient fluorescent probes, and a deeper, theoretical understanding of fundamental processes. Frequently, fluorescence techniques are tailored to attack a specific biological problem. These new methods in turn produce new physical situations and phenomena which are often of interest to the physical chemist. Thus, progress in one area stimulates renewed interest in other areas. The goal of this book is to provide detailed monographs on the use of fluorescence to investigate problems at the forefront of biochemistry and cell biology. This book is not meant to be a comprehensive survey but rather to highlight areas of recent developments. It is designed to be readable to the novice and yet provide sufficient detail for the expert to keep abreast of recent developments.

The book is organized so that it proceeds from simple biochemical systems to more complex cell biological ones. Chapter 1 on fluorescence quenching of biological structures is a good introductory chapter. It introduces a number of elementary concepts and discusses applications to proteins and biomembranes. This chapter reveals the relative ease of obtaining structural dynamic information using fluorescence quenching. The role of divalent cations in biological systems has attracted considerable recent attention. Chapter 2 presents the properties of fluorescent analogues to divalent cations, the lanthanides. It discusses in detail how water coordination and cation binding site heterogeneity can be determined in biological systems. The controversial use of lanthanides for obtaining distance information by fluorescence energy transfer is also critically reviewed. In Chapter 3 fluorescence applications to a new technology, on-line, real time DNA sequencing —is discussed. The synthesis and sequencing reactions of fluorescently labeled nucleic acid primers are examined. The development of real time sequencing capabilities with multiple fluorescent primers offers an exciting and potentially powerful alternative to conventional techniques. Chapter 4 provides an excellent introduction to the use of fluorescence anisotropy for the study of macromolecular motion. A description of both time-resolved and modulation instrumentation and data analysis is presented. Applications to molecular motions of tRNA and elongation factor Tu are discussed. Chapter 5 is another good general chapter that demonstrates application of a variety of fluorescence techniques to increasingly complex systems. It explores the use of fluorescence energy transfer and fluorescence quenching to study receptor–G protein interactions. The focus is more on the system and this chapter demonstrates how both intrinsic and extrinsic fluorescent probes can be utilized to attack biochemical problems. Chapter 6 is along similar lines as Chapter 5. It concentrates on antibody–receptor interactions. Again, fluorescence spectroscopy is the tool used for characterizing protein–protein interactions. The kinetics and aggregation of receptors are studied with the goal of developing quantitative models. The biological membrane is a crucial component in these more complicated systems and fluorescence applications to membrane systems are covered in more detail in Chapter 7. This chapter describes the use of fluorescence energy transfer to obtain structural and dynamic information on proteins in artificial membrane systems. Chapters 8 and 9 proceed along this line to cellular systems. Chapter 8 discusses fluorescence energy transfer and photobleaching on cell surfaces. The energy transfer experiments represent a new application of video imaging of cell surfaces. A significant portion of this chapter deals with the technology of fluorescence video imaging of cells. Chapter 9 is a good background chapter on fluorescence photobleaching. It also discusses several cell biological problems involving membrane organization and dynamics. Experimental work on a variety of membrane transport processes is examined.

There are several themes which loosely run through this book. One is the similarity of techniques used to attack biological problems. The chemistry of fluorescent labeling has common features in a diversity of systems. Fluorescence quenching, energy transfer, and anisotropy appear in different contexts as techniques for obtaining structural and dynamic information. Although in many cases complex biochemical systems are studied, a significant quantitative rigor is involved. This persists from data analysis to theoretical models. These common features assist interactions between investigators who work on very different systems. It is hoped that this collection of monographs reveals the excitement and the diversity of skills and approaches used in current research.

T. Gregory Dewey

Denver

Contents

Chapter 1

Fluorescence Quenching Reactions: Probing Biological Macromolecular Structures
Maurice R. Eftink

Chapter 2
Luminescent Trivalent Lanthanides in Studies of Cation Binding Sites
Matthew Petersheim

Chapter 3
Continuous, On-Line, Real Time DNA Sequencing Using Multifluorescently Tagged Primers
John Brumbaugh, Lyle Middendorf, Dan Grone, and Jerry Ruth

Chapter 4
Time-Resolved Fluorescence in Biology and Biochemistry

David M. Jameson and Theodore L. Hazlett

Chapter 5
Fluorescence Investigations of Receptor-Mediated Processes
William J. Phillips and Richard A. Cerione

Chapter 6
Analysis of Ligand Binding and Cross-Linking of Receptors in Solution and on Cell Surfaces: Immunoglobulin E as a Model Receptor
Jon Erickson, Richard Posner, Byron Goldstein, David Holowka, and Barbara Baird

Chapter 7
Fluorescence Energy Transfer in Membrane Biochemistry
T. Gregory Dewey

Chapter 8
Imaging Membrane Organization and Dynamics
Richard A. Cardullo, Robert M. Mungovan,
and David E. Wolf

Chapter 9

The Dynamic Parameter: Fluorescence
Photobleaching as a Tool to Dissect Space in
Biological Systems

Melvin Schindler, Paramjit K. Gharyal, and Lian-Wei Jiang

Chapter 1

Fluorescence Quenching Reactions
Probing Biological Macromolecular Structures

Maurice R. Eftink

1. INTRODUCTION

The quenching of the fluorescence of biomacromolecules by solute quenchers has become a widely used and powerful technique (Lehrer, 1976; Lehrer and Leavis, 1989; Lakowicz, 1983; Eftink and Ghiron, 1981; Eftink, 1991). Such quenching reactions have been used primarily to obtain topographical information about proteins, nucleic acids, and membrane systems. The accessibility of intrinsic or extrinsic fluorescence probes (e.g., the amino acid, tryptophan), which are attached to a biomacromolecule, to small quenchers (e.g., iodide, acrylamide, oxygen) is directly determined by such reactions. Conformational changes in the biomacromolecule can then be monitored in terms of changes in the accessibility to the quencher of the fluorophore. In addition to such topographical information, in some cases information about the conformational dynamics of globular proteins has been obtained with solute quenching reactions.

The popularity of fluorescence quenching reactions stems not only from the usefulness of the information they report, but also from the ease of performing quenching experiments. Equipped with little more than a fluorometer, cuvettes, pipettes, and common reagents, an investigator can embark upon a quenching experience.

MAURICE R. EFTINK • Department of Chemistry, University of Mississippi, University, Mississippi 38677.

In this chapter I will give an overview of the method, with particular emphasis on practical matters and with examples of how the method can give meaningful information about various biological assemblies. My personal interest lies more toward the mechanisms of quenching reactions and the use of these reactions to study protein dynamics. In a separate, recent review (Eftink, 1991), I have dealt with topics such as the efficiency of quenching, static quenching and transient diffusional effects, and quencher penetration and unfolding models of protein dynamics. I will downplay these molecular biophysical topics here and will instead emphasize the uses of solute quenching reactions to obtain topographical information.

Two of the following sections will deal with experimental details and the analysis of quenching data. Another section will describe several experimental variables that can be used to further exploit quenching reactions. In the last section I will present selected examples of the use of solute quenching reactions to study biomacromolecular conformations and conformational changes, with systems ranging in complexity from peptides, to membranes, to nucleoprotein assemblies. In this chapter I do not intend to give an exhaustive review of solute quenching; rather I hope to highlight a variety of applications and strategies of the method so as to challenge the occasional practitioner and to provide an introduction to the new practitioner.

2. BASIC CONCEPTS

2.1. The Stern–Volmer Equation

The quenching of an excited singlet state by a solute quencher can be described by the following general reaction scheme:

Here A is a chromophore, which, upon excitation to a higher electronic energy level (A*) by absorption of a photon, can emit a photon (fluorescence) with radiative rate constant k_r. The excited state A* may also return to the ground state by other nonradiative mechanisms, which we lump together as the nonradiative rate constant, k_{nr}. The fluorescence lifetime, τ_0, and quantum yield, Φ_0, of A* are defined as (the subscript "0" indicates the absence of solute quencher)

$$\tau_0 = 1/(k_r + k_{nr}) \tag{1}$$

$$\Phi_0 = \frac{k_r}{k_r + k_{nr}} = \frac{\tau_0}{\tau_N} \tag{2}$$

where $\tau_N = 1/k_r$.

In the presence of the solute quencher, Q, additional reactions are possible which lead to a more rapid return of excited states back to ground states. The excited A* may collide with Q, with diffusional rate constant k_d, to form an excited-state encounter complex, $(A \cdots Q)^*$. If Q is in fact a quencher, then such $(A \cdots Q)^*$ complexes will undergo an internal fluorescence quenching reaction, with rate constant k_i. The mechanism of this internal quenching reaction is not known for certainty for many quenchers. The mechanism may involve electron spin exchange (i.e., paramagnetic quenchers and oxygen) or spin orbital coupling (i.e., halides and other heavy atoms) to enhance intersystem crossing, or electron transfer (i.e., amides). (In this review I will consider only those quenching reactions that occur at a short distance; long-range resonance energy transfer is not considered.) Regardless of the mechanism, the kinetics of contact solute quenching reactions are essentially the same and the reactions are useful for biochemists, as described below.

The quencher may also form a ground state complex, $(A \cdots Q)$, by encountering the ground state of A. Such a complex may also absorb a photon to directly populate the excited-state encounter complex, $(A \cdots Q)^*$. The internal quenching reaction may then proceed immediately after excitation. In Scheme I, k_r and k_{nr} are assumed to be the same for both A* and $(A \cdots Q)^*$; also k_d and k_{-d} are assumed to be the same for ground-state and excited-state diffusional reactions. (These are reasonable and common assumptions, but a more elaborate Scheme can be proposed.)

As indicated by the labels in Scheme I, the left route, in which the diffusional reaction between A* and Q is usually the limiting step for quenching, is described as being a dynamic process. The right route, in which a ground-state encounter complex is formed before excitation, is referred to as static quenching. This reference is due to the fact that the latter quenching mechanism is not limited by a diffusional step and thus should not be viscosity dependent. If Q is an efficient quencher (i.e., k_i is large compared to k_{nr} and k_r), then the $(A \cdots Q)^*$ complex will be deactivated virtually as soon as it is formed by either route.

In the presence of a quencher, the quantum yield, including excitation into both A and $(A \cdots Q)$, will be

$$\Phi = \frac{k_r(1/\tau_0 + k_i + k_{-d} + k_d[Q] + K_a[Q](1/\tau_0 + k_d[Q] + k_{-d}))}{(1 + K_a[Q])(1/\tau_0 + k_i + k_{-d} + k_d[Q] + k_i k_d[Q]/\tau_0)} \quad (3)$$

where $K_a = k_d/k_{-d}$.

The ratio of the quantum yield in the absence and presence of Q is

$$\Phi_0/\Phi = 1 + \gamma k_d \tau_0[Q] + \gamma K_a[Q] + \gamma k_d \tau_0 K_a[Q]^2 \quad (4a)$$

$$= (1 + \gamma k_d \tau_0[Q])(1 + \gamma K_a[Q]) \quad (4b)$$

where γ is the efficiency of the quenching process and is defined as (Eftink and Ghiron, 1981)

$$\gamma = \frac{k_i}{k_i + 1/\tau_0 + k_{-d} + k_d[Q]} \quad (5)$$

If the internal quenching rate constant is much larger that the other rate constants, the $\gamma = 1.0$ and the quencher would be described as being 100% efficient or as being a strong quencher. For such an efficient quencher, Eq. (4) becomes the following, which is the Stern–Volmer equation:

$$\Phi_0/\Phi = (1 + k_q \tau_0[Q])(1 + K_a[Q]) \quad (6a)$$

$$= (1 + K_{sv}[Q])(1 + K_a[Q]) \quad (6b)$$

Here k_q, the quenching rate constant, is equal to γk_d (and $k_q = k_d$ if $\gamma = 1.0$), and K_{sv}, the dynamic quenching constant, is equal to $k_q \tau_0$. K_a is referred to as the static quenching constant. The total degree of quenching is thus a product of dynamic, $(1 + K_{sv}[Q])$, and static, $(1 + K_a[Q])$, terms. (The static term is usually smaller than the dynamic term, but there are cases in which static quenching predominates.)

According to this relationship, plots of Φ_0/Φ versus [Q] should be upward curving. Shown in Fig. 1A is an example of such a plot, as well as examples of other forms of the Stern–Volmer equation to be described below. This upward-curving behavior is commonly observed for efficient quenchers. As has been explained elsewhere (Eftink *et al.,* 1987), if the quenching efficiency is much less than unity (i.e., k_i is not greater than k_{nr}, k_r, and k_{-d}), this will result in Stern–Volmer plots that do not obey Eq. (6). In fact, an inefficient quencher may give a Stern–Volmer plot that appears linear or even downward-curving, so caution should be used when employing inefficient quenchers. The upward curvature predicted by Eq. (6) is sometimes not apparent if high Q concentrations are not employed. That is, the $[Q]^2$ term in Eq. (6) may be so small, at low [Q], that an upward curvature is not seen. Also, as discussed below, when there is heterogeneity in the fluorescence of a sample (i.e., more than one type of fluorophore), the

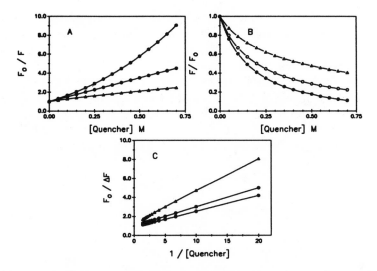

FIGURE 1. Simulated plots of fluorescence quenching data. (A) Stern–Volmer plot of F_o/F versus [Q]. ●, plot of Eq. (6) with K_{sv} = 5 M^{-1} and K_a = 1.0 M^{-1}; ○, with K_{sv} = 5 M^{-1} and no static quenching; ▲, heterogeneous system with $K_{sv,1}$ = 5.0 M^{-1}, $K_{sv,2}$ = 1.0 M^{-1}, f_1 = 0.5, and no static quenching. (B) Direct plots of F/F_o versus [Q] for the three cases above. (C) Plot of $F_o/\Delta F$ versus 1/[Q] for the above three cases. Notice how this plot expands the data at low [Q] and compresses the data at high [Q].

Stern–Volmer relationship becomes more complicated and downward-curving plots are predicted.

The following alternate form of the Stern–Volmer relationship is often used:

$$\Phi_0/\Phi = (1 + K_{sv}[Q])\exp(V[Q]) \tag{7}$$

In Eq. (7), the static term is given by $\exp(V[Q])$, instead of $(1 + K_a[Q])$. Of course these factors are approximately the same, since $\exp(x) \approx 1 + x$, when x is small. A simple explanation for the different static terms is that the $(1 + K_a[Q])$ term derived from Scheme I is for the formation of a one-to-one $(A \cdots Q)$ ground-state complex. If higher-order ground-state complexes [i.e., $(A \cdots Q_2)$, $(A \cdots Q_3)$] can form, then the $\exp(V[Q])$ term is more appropriate. As has been discussed elsewhere (Eftink and Ghiron, 1976a), the static constant, V, can be considered to be an active element surrounding the A molecule. If one or more Q molecules happen to be within this volume element at the instant of photon absorption, then instantaneous (static) quenching occurs. The magnitude of V can be related to an active radius, R, by the relationship, $V = (4/3)N'\pi R^3$.

Fluorescence quenching measurements are usually performed by observing changes in the steady-state fluorescence quantum yield (or intensity, F) as quencher is added. Quenching can also be observed by measuring fluorescence lifetimes. In the presence of quencher, any $(A \cdots Q)^*$ states, which are formed by direct excitation into ground-state $(A \cdots Q)$ complexes, will usually be quenched instantaneously and completely. This static process will not result in a change in the observed lifetime. Those A^* molecules that are quenched by the dynamic route will have an apparent lifetime, in the presence of quencher, given by the following relationship:

$$\tau = \frac{1}{k_{nr} + k_r + k_q[Q]} \tag{8}$$

The lifetime Stern–Volmer relationship for the dynamic quenching process will be[†]

$$\tau_0/\tau = 1 + K_{sv}[Q] \tag{9}$$

Fluorescence lifetime measurements require more sophisticated equipment, but if such measurements can be made, then in principle dynamic and static quenching processes can be separated by a direct comparison of lifetime and intensity quenching data. Equations (6) and (9) predict that static quenching contributes only to the intensity data.

The parameter of interest, of course, in fluorescence quenching experiments is the quenching rate constant, k_q. In studies with biomacromolecules, the k_q value is a kinetic measure of the exposure of an attached fluorophore. For the quenching of a free fluorophore, such as indole in water, by an efficient quencher, the k_q value should be equal to the diffusion-limited rate constant, k_d. From the Smoluchowski theory of diffusion-limited reactions, the value of k_d for a biomolecular reaction can be calculated as

$$k_d = 4\pi D R_0 N' \tag{10}$$

where D is the combined diffusion coefficients of the quencher and fluorophore, R_0 is the effective encounter distance (usually slightly larger than the sum of the molecular radii), and N' is 6.02×10^{20}. For typical values of $D = 1.5 \times 10^{-5}$ cm^2/sec (in water at 25°C) and $R_0 = 6.5$ A, a k_d value of about 7 $\times 10^9$ M^{-1} sec^{-1} is calculated. Thus, for an efficient quencher, a k_q value of this magnitude is expected for the quenching of a free fluorophore. Quenchers or fluorophores that have a larger radius also have a smaller

---

† I will not consider transient effects (Lakowicz et al., 1986, 1987; Nemzek and Ware, 1975; and Andre et al., 1978) or more complicated kinetic expressions that arise due to the multistep nature of Scheme I (Gratton et al., 1984). For a discussion of such matters, see Eftink (1991) and the above references. Here we present the basic Stern–Volmer relationships, which apply reasonably well to most studies.

diffusion coefficient, so this same k_d value is a rough estimate for many reactant pairs (provided the quencher and/or fluorophore is fully reactive at all points on the molecule). An estimate of the efficiency of a quencher can be made as $\gamma = k_q/k_d$.

Steric or electrostatic factors may reduce the value of k_q; the existence of such steric or electrostatic shielding is revealed by quenching studies with biomacromolecular assemblies. For example, for the acrylamide quenching of Trp residues in globular proteins, a k_q range of ~ 0.01 to 4×10^9 M^{-1} sec^{-1} has been found (Eftink and Ghiron, 1976b; Eftink, 1991). Smaller k_q values indicate that Trp residues are sterically buried within a globular protein; larger k_q values indicate that a Trp residue lies on the surface of a protein. As shown theoretically by Johnson and Yguerabide (1985), the k_q for a fully exposed fluorophore, attached to a macromolecule, is expected to be reduced, by roughly a factor of two, from the above value for free reactants. (See Section 5.1.)

A useful equation to estimate the diffusion coefficient for a quenching reaction is the Stokes–Einstein relationship

$$D_i = k_b T / 6 \pi \eta R_i \tag{11}$$

where k_b is the Boltzmann constant, η is the viscosity of the medium, and R_i is the molecular radius of an individual reactant. The important feature of this relationship, together with Eq. (10), is that k_q is predicted to vary with T/η for a dynamic quenching process. This general behavior has been observed for several quencher–fluorophore systems, but careful studies have revealed deviations, which appear to be due to either the limitations of the Stokes–Einstein theory (Alwattar *et al.*, 1973; Eftink and Ghiron, 1987a) or the inefficiency of the reaction (Olea and Thomas, 1988; Eftink and Ghiron, 1976a).

2.2. Heterogeneous Emission

For the common cases in which a biomacromolecule possesses two or more different fluorescing centers, the Stern–Volmer equation, for intensity quenching, has the following form (substituting fluorescence intensity, F, for quantum yield, Φ)

$$\frac{F_0}{F} = \left(\sum_{i=1}^{n} \frac{f_i}{(1 + K_{sv,i}[Q])(1 + K_{a,i}[Q])} \right)^{-1} \tag{12}$$

or

$$\frac{F}{F_0} = \sum_{i=1}^{n} \frac{f_i}{(1 + K_{sv,i}[Q])(1 + K_{a,i}[Q])} \tag{13}$$

If static quenching is neglected, the latter equation becomes

$$\frac{F}{F_0} = \sum_{i=1}^{n} \frac{f_i}{1 + K_{sv,i}[Q]} \tag{14}$$

In these expressions, $K_{sv,i}$ and $K_{a,i}$ are the dynamic and static quenching constants for component i, f_i is the fractional intensity (at the excitation and emission wavelengths used) of component i, and n is the number of components. In Eqs. (12) and (13), $\exp(V_i[Q]$ may be used instead of (1 + $K_{a,i}[Q]$). Stern–Volmer plots of F_0/F versus [Q] will be downward curving for such heterogeneous systems, so long as the $K_{sv,i}$, for the components, differ by a factor of four or more (see Fig. 1A). Static components will produce upward-curving tendencies, which can mask any downward curvature. The fluorescence lifetimes of each component will, in principle, follow Eq. (9).

Simulations of the effect of heterogeneous emission on Stern–Volmer plots have been presented (Lehrer and Leavis, 1978; Eftink and Ghiron, 1981). As discussed in a following section, a problem in data analysis is to recover the individual quenching constants for a heterogeneously emitting system.

2.3. Partitioning among Subphases

In some systems, particularly those involving membranes or micelles, there is the possibility that the quencher may be partitioned between the aqueous phase and the above types of subphases. The result of this phenomenon is that the local concentration of quencher in the subphase may be much different than the overall concentration of the solution. For example, apolar quencher molecules may favorably partition into the hydrocarbon-like core of a membrane vesicle, thus giving a high local concentration of the quencher within the vesicle bilayer. If the fluorophore is also located within the subphase, then enhanced quenching may occur. (Also, of course, if there is extensive depletion of the apolar quencher from the aqueous phase, there may be reduced quenching of fluorophores in the aqueous phase; this will depend on the partition coefficient and the ratio of the volumes of the subphase and aqueous phase.)

Blatt and Sawyer (1985) and Blatt et al. (1986) have presented thorough discussions of the relationships that can be used to describe solute quenching reactions in such compartmentalized systems. For the usual case in which the fluorophore is associated with the subphase and the quenching reaction occurs only with quencher molecules that are also in the subphase, the Stern–Volmer equation (7) will still apply, but the pertinent value for the quencher concentration will be the local concentration in the subphase, $[Q]_2$ (here I use the label "1" for the main aqueous phase, and "2" for the subphase). When the correction in concentration is made, the quenching rate constant, $k_{q,2}$, will be that for the collisional reaction in the subphase. This $k_{q,2}$ and the static quenching constant, V_2, are the parameters of interest in describing the quenching process. The value of $[Q]_2$ is generally unknown, however. It can be related to a partition coefficient, $K_P = [Q]_2/[Q]_1$. From the conservation of mass relationship [Eq. (15)] one can derive the following expression [Eq. (16)] for $[Q]_2$ as a function of the total quencher concentration, $[Q]_T$, and the volumes of the subphase, V_2, and the total volume of the solution, V_T:

$$[Q]_T V_T = [Q]_2 V_2 + [Q]_1 V_1 \qquad (15)$$

$$[Q]_2 = \frac{K_P[Q]_T}{1 + K_P V_2/V_T - V_2/V_T} \qquad (16)$$

By substituting this expression for $[Q]_2$ into the Stern–Volmer equation (7), the following general relationship obtains for solute quenching in a compartmentalized system:

$$\frac{F_0}{F} = \left(1 + \frac{k_{q,2}\tau_0 K_P[Q]_T}{1 + K_P V_2/V_T - V_2/V_T}\right)\exp\left(\frac{V K_P[Q]_T}{1 + K_P V_2/V_T - V_2/V_T}\right) \qquad (17)$$

By measuring fluorescence quenching profiles, as a function of the total quencher concentration, for several ratios of V_2/V_T, one can in principle fit this equation to the data sets with three parameters, $k_{q,2}$, K_P, and V. This approach has been used by Lakowicz et al. (1977) to study quenching in phospholipid vesicles and by Encinas and Lissi (1982) and Blatt and Sawyer (1985) to study quenching reactions in micelles. The latter group has further considered the possibility that the interaction of the quencher with the subphase may be more appropriately described as a saturable binding process, instead of a partitioning process.

3. EXPERIMENTAL DETAILS

Most solute quenching experiments are very easy to perform and "first time" experimenters usually have success. In this section I give, in no particular order, some practical advice. The chapter by Lehrer and Leavis (1978) also contains many experimental details.

The minimum requirements for such experiments are a fluorometer, cuvettes, pipettes, and fluorophore and quencher solutions. It is preferable to have both excitation and emission monochromators, so that the desired fluorescence states can be photoselected. For example, by using the excitation wavelength range of 290–300 nm, one can selectively excite Trp residues in a protein, without dual excitement of Tyr residues. It is wise to select an emission wavelength where Rayleigh and Raman scattering are minimized (i.e., with 295-nm excitation, a Raman peak occurs at about 328 nm in water). A convenient way to add quencher to a fluorescent solution in a cuvette is by simply adding aliquots of a concentrated (i.e., 8 M for acrylamide, 5 M for KI) quencher solution. This can be done with a simple syringe or micropipette (positive displacement micropipettes work well), or with an automatic syringe or peristaltic pump. It is advised to have a second cuvette, which contains the buffer without the fluorescing species. This blank solution can be used to measure the baseline, and, by adding the quencher to this blank, one can check whether the quencher solution has a fluorescent impurity. An alternate way to add the quencher to a fluorescing solution is by the Job method of continuous variation. Here one starts with two solutions, the first containing the fluorophore and the second containing the fluorophore (at the same concentration) plus the quencher. By measuring the fluorescence of each solution, and then transferring aliquots of the second to the first solution, a quenching profile can be obtained. This method is particularly useful in cases in which the quencher is a salt and it is desired to maintain a constant ionic strength. For example, the first and second solutions can be prepared to contain the same concentration (i.e., 0.5 M) of KCl and KI, respectively. When these solutions are mixed, the ionic strength will remain the same. With either the Job or direct addition methods, thorough mixing of the solution is of course required. This can be done by a magnetic flea, by aspiration with a disposable pipette, or any other trusted procedure.

Temperature and pH should ordinarily be controlled by circulating baths and buffers. As mentioned above, when using a charged quencher, like KI, it is best to work with a constant ionic strength, particularly if electrostatic interactions between the biomacromolecule and quencher are to be expected. Lehrer (1976) demonstrated the problems that can arise, if ionic strength is not maintained, by a study of the iodide quenching of human serum albumin. Iodide appears to interact weakly with this protein, resulting in downward-curving Stern–Volmer plots, even though there is a single Trp residue. By maintaining a high ionic strength (with KCl), and particularly by adding KSCN (apparently SCN⁻ competes with I⁻ for a weak binding site on the protein), the Stern–Volmer plots were found to be more linear. In Fig. 2 we demonstrate a similar effect with the KI quenching of the single Trp in staphylococcal nuclease. When KI is added directly to the protein in a low-ionic-strength buffer (i.e., not maintaining a constant ionic strength, but allowing it to increase as KI is added), a downward-curving Stern–Volmer plot is obtained. Yet, when the same experiment is done with the ionic

FIGURE 2. Stern–Volmer plot for the KI quenching of staphylococcal nuclease at pH 5.0. $\bigcirc$, ionic strength not maintained and no salt present when $[Q] = 0$. $\bullet$, ionic strength maintained at 0.95 M by addition of KCl. Other conditions: 0.01 M Na acetate buffer, 20°C. A linear Stern–Volmer plot is found at constant ionic strength. When the ionic strength is varied, the plot is downward-curving and analysis with Eq. (22) (i.e., assuming $K_{sv,2}$ = 0) yields $K_{sv,1}$ = 13.5 M^{-1} and

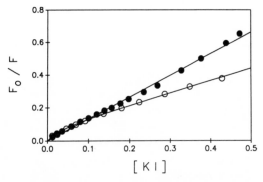

$F_1 = 0.185$. If one drops the assumption that $K_{sv,2} = 0$ and uses Eq. (13), fits are obtained with values of $K_{sv,1}$ = 9.5 M^{-1}, $K_{sv,2}$ = 0.54 M^{-1}, and f_1 = 0.149, even though there is a single Trp in this protein (the latter fit is shown as the solid line through these data). When ionic strength is maintained, K_{sv} = 1.32 M^{-1}.

strength maintained at 1.0 M, a linear Stern–Volmer plot is obtained. As I will stress in Section 5.3, one must be cautious in interpreting nonlinear Stern–Volmer plots for charged quenchers if one has not used a constant ionic strength.

It is usually sufficient to measure the quenching of the fluorescence at a single emission wavelength. The independent variable in Eq. (6) is actually the quantum yield; measurement of the quantum yield requires integration over the entire spectrum. If the fluorescence is homogeneous, then measurement of the fluorescence intensity, F, at a single wavelength (usually the maximum) is sufficient, since $F \propto \Phi$. When there is ground-state heterogeneity (e.g., more than one fluorescing Trp residue), then Eq. 13 applies. If the fluorescence intensity, F_λ, is measured at a particular wavelength, λ, then the fractional contributions, $f_{i,\lambda}$, may also depend on wavelength. By determining $f_{i,\lambda}$ at several λ, the spectral contours of the components can be determined (see below). It is, of course, tedious to measure quenching profiles at several wavelengths. We foresee that the use of multichannel, diode array detectors will greatly facilitate data acquisition at multiple wavelengths.

In a previous article (Eftink and Ghiron, 1981) we listed several solute quenchers and their characteristics. Table I is an update of this information. An approximate value for the quenching efficiency, γ, of each quencher, with Trp or indole as fluorophore, is given, as well as selected information on the efficiency of quenching of other fluorophores. It is desirable to use a quencher-fluorophore system that has an efficiency near unity. If γ is much less than unity, higher quencher concentrations will be required and anomalous temperature and viscosity dependencies may be observed (Eftink and Ghiron, 1981). Acrylamide is a useful quencher of Trp in proteins, but, as

TABLE I. Quenchers of Indole Fluorescence

Quencher	k_q ($\times 10^{-9}$ M^{-1} sec^{-1})	V (M^{-1})	γ	Comments
Neutral				
Oxygen[a]	12.3		1.0	Nonpolar, special cell required
Acrylamide	7.1	2.5	1.0	Polar, $\epsilon_{295} = 0.25$ M^{-1} cm^{-1}
Succinimide	4.8	1.0	0.7	Polar, $\epsilon_{295} = 0.03$ M^{-1} cm^{-1}
Trichloroethanol	5.0	2.0	0.8	Nonpolar, denatures and/or precipitates proteins at high concentrations
Trifluoroacetamide[b]	1.9	0.34	0.3	Polar
Dichloroacetamide[c]	6.9	1.5	1.0	
Methionine	0.7	—	0.1	
β-Mercaptoethanol[d]	~3	—	0.5	
Hydrogen peroxide[e]	4.5	—	0.7	Polar, $\epsilon_{300} = 0.9$ M^{-1} cm^{-1}, reacts with side chains of proteins
Anions				
I$^-$	6.4	2.0	1.0	
Br$^-$	0.2	—	0.04	
IO$_3^{-}$[f]	4	—	0.7	
NO$_3^{-}$[f]	11.0	—	1.0	Large absorbance above 300 nm. Also NO$_2^-$
Cations				
Pyridinium-HCl	9.4	5.0	1.0	Protonated only below pH 5.3, N-methylpyridinium also possible, $\epsilon_{295} = 0.42$ M^{-1} cm^{-1}
Imidazolium-HCl	3.6	—	0.6	$\epsilon_{295} = 0.05$ M^{-1} cm^{-1}
Tl$^+$	9.2	2.1	1.0	Must use acetate, formate, or fluoride salt; *poisonous*
Eu^{3+}[g]	6.0	2.5	1.0	
Ag$^+$[h]	6.1	—	1.0	Forms ground-state complex with indole; forms mercaptides
Cs$^+$	1.1	—	0.2	

Most values were determined in our laboratory and apply to room temperature, neutral aqueous buffers unless noted. Other values were obtained from the following references: [a] Lakowicz and Weber (1973a); [b] Midoux et al. (1984); [c] Froehlich and Nelson (1978); [d] Bushueva et al. (1975); [e] Cavatorta et al. (1979); [f] Altekar (1977); [g] Ricci and Kilichowski (1974); [h] Chen (1973). The efficiency value is determined as $\gamma = k_q/k_d$ where k_d is calculated from Eq. (10). In general, a value of $k_d \approx 6.5 \times 10^9$ M^{-1} sec^{-1} is used as an estimate for all quenchers. If $k_q/k_d > 1.0$, we report only a value of 1.0; either the effective D_i or R_o is larger than expected in these cases, or some long-range reaction is possible.

TABLE II. Acrylamide, Succinimide, Iodide, and Oxygen Quenching Constants for a Number of Fluorescence Probes in Water

Fluorophore	τ_0 (nsec)	k_q^A	k_q^S	k_q^I	k_q^0
		($\times 10^{-9}$ M^{-1} sec^{-1})			
Tyrosine	3.2	7.0	5.3	—	12[a]
NADH	0.4	2.1	0	0	
Riboflavin	4.8	0.2	0.1	9.9	4.0[a]
Proflavin	5.0	0, 2.0[d]	0	12.7	
Fluorescein	4.2	0	0.2	1.5	
Naphthaleneacetic acid	28.2	4.6	0.83	3.7	9.3[c]
Anthracene-9-carboxylic acid	1.5	2.4	0.1	—	7.7[a]
PRODAN	1.5	0	0	2.8	
IAEDANS	12	1.2	0.1	—	
DENS	30.3	3.2	0.1	0.03	7.7[a]
1,N-ethanoadenosine	25.0	2.5	0	7.8	
Umbelliferone	5.5	2.1	0	2.5	
Eosin Y	1.3	0	0	2.2	
β-Methylumbelliferone	5.4	3.2	0	2.8	
TMA-DPH	0.3	0	0	—	
Pyrenebutyric acid	97.5	3.4	0.06	—	10.0[b], 9.6[c]
Carbazole	10.4	7.0	5.4	—	

Most values taken from Eftink *et al.* (1987). Other values taken from the following references: [a] Lakowicz and Weber (1973a); [b] Vaughn and Weber (1970); [c] Eftink and Ghiron (1987a); [d] Zinger and Geacintov (1988).

shown in Table II, its efficiency for quenching other common biochemical fluorophores is less than unity. Molecular oxygen appears to be an efficient quencher of virtually all fluorophores, but a special high-pressure cell (Lakowicz and Weber, 1973a) is needed for its use, except for those fluorophores, such as pyrene, that have a very long τ_0 (Vaughn and Weber, 1970). Iodide is the anionic quencher of choice, as it has a reasonably high efficiency for many fluorophores (Lehrer, 1971; Eftink *et al.*, 1987). A corresponding cationic quencher is not well established. Thallous ion is an efficient quencher (Ando and Asai, 1980), but it must be used with caution (poisonous). Also, the solubility of chloride, bromide, iodide, and phosphate salts of Tl$^+$ is very low (15 mM for TlCl); these anions must be avoided in buffers. Acetate, formate, fluoride, perchlorate, carbonate salts and buffers can be used with Tl$^+$. This quencher also seems to cause certain proteins to precipitate.

Listings of several less efficient quenchers have been given by Steiner and Kirby (1969), Bushueva *et al.* (1975), Altekar (1977), and Bohorquez *et al.*, (1984). Quenchers listed in Table I seem to quench primarily by a close-range interaction (collision). Quenchers that probably act by long-range energy transfer are not included in Table I, although NO$_3^-$ and NO$_2^-$ have significant absorbance above 300 nm and may quench, at least partially, by this mechanism.

A special class of quenchers are those in which a quenching moiety, such as a nitroxide or bromo group, is part of the alkyl chain of a fatty acid or phospholipid (Thulborn and Sawyer, 1978; Bieri and Wallach, 1975; Markello et al., 1985). These quencher-lipids are extremely useful for studying the degree of membrane penetration of extrinsic fluorescent probes and Trp residues of peptides and proteins (Stubbs and Williams, 1991; Eftink, 1991; Chattopadhyay and London, 1987).

Since quencher concentrations as high as 0.5–1.0 M are sometimes added, it is necessary that the quencher be very pure. Reagent-grade acrylamide and succinimide must be recrystallized. Clean acrylamide should have a molar extinction coefficient of about 0.25 M^{-1} cm^{-1} at 295 nm and no absorbance above 310 nm. Recrystallization of succinimide removes some acidic impurity. Iodide solutions have a tendency to form I_3^- and turn yellow. This can be avoided by adding small amounts ($\sim 1 \times 10^{-4}$ M) of $Na_2S_2O_3$ to stock KI solutions. Spin-labeled fatty acids and phospholipids may have a significant portion of nitroxide groups that are not paramagnetic. The ratio of spins per molecule can be experimentally determined as described by Chattopadhyay and London (1987).

Fluorescence data should be corrected for absorption of light (screening) by the quencher at the excitation (and emission, if necessary) wavelength. This absorptive screening correction is generally taken to be the factor antilog ($l\epsilon_\lambda[Q]/2$), where ϵ_λ is the molar extinction coefficient of the quencher at the excitation λ, and l is the pathlength of the cuvette. (This correction applies for a right-angle geometry.) In cases where the absorptive screening is very large (i.e., $\epsilon_\lambda[Q] > 0.3$ or so for a 1-cm path), reduced-pathlength cuvettes can be used. Alternatively, front-face detection or angled cuvettes may be used. It should be mentioned that ϵ_λ values may be temperature and solvent dependent, as is the case for acrylamide. Also we have found that the ϵ_λ at 295 nm for acrylamide is larger in the presence of I^- than Cl^- salts; this may indicate a weak charge transfer interaction between acrylamide and I^- (Eftink and Selvidge, 1982).

In some cases there may be a chemical or physical interaction between the solute quencher and the biomacromolecule and such interactions should be either avoided or appreciated. The electrostatic interaction between charged quenchers and proteins was mentioned above. Neutral quenchers may interact specifically or nonspecifically with certain proteins; a few cases are known in which acrylamide binds specifically to proteins (Narasimhulu, 1988; Eftink and Ghiron, 1987b). The apolar quencher, trichloroethanol (TCE), seems to be attracted to apolar regions in proteins (Eftink et al., 1977). In fact, the addition of TCE (0.2 M or above) to some proteins appears to cause their unfolding, as evidenced by a dramatic upward curvature of Stern–Volmer plots. This effect is similar to the action of other apolar solutes (e.g., 2-chloroethanol) on protein structures. In general, if an increase in [Q] causes an abrupt upward curvature in a Stern–Volmer plot, one should suspect an induced change in the conformation of the biomacromolecule.

When membrane or micellar systems are studied, apolar quenchers will have a tendency to be partitioned into the lipid subphase. This phenomenon is the basis for the quenching of certain membrane-incorporated fluorophores by quenchers that also partition into the membrane (Blatt and Sawyer, 1985; Blatt *et al.,* 1986). The key for such applications is to determine the effective concentration of the quencher within the lipid phase, so that meaningful biomolecular rate constants can be evaluated (see Section 2.3).

Chemical reaction between the fluorophore and quencher will usually not be a problem, but there are a few known reactions. TCE appears to promote a photochemical modification of Trp residues in proteins. This reaction was successfully employed by Toulmé *et al.* (1984) to selectively modify certain Trp residues in the gene 32 protein from phage T4. Acrylamide will slowly react with primary amino groups and sulfhydryl groups at high pH (Geisthardt and Kruppa, 1987; Hashimoto and Albridge, 1970; Danileviciute *et al.,* 1981). These reactions are probably too slow at neutral pH and room temperature to be of consequence and it is not clear if the adducts that form are still capable of quenching fluorescence. Mercuric ions, Hg^{2+} (Sluyterman and DeGraaf, 1970), and silver ions, Ag^{2+}, form mercaptides with sulfhydryl groups (Chen, 1973). Hydrogen peroxide reacts with various amino acid side chains (Cavatorta *et al.,* 1979).

It is good practice to try to determine whether a quencher reacts with or perturbs the conformation of a protein or other biomacromolecale. With enzymes, the most convenient test is to determine if the catalytic activity is altered by the presence of the quencher (one should perform assays with a substrate concentration below K_m, so that the second-order catalytic rate constant is compared). Other tests might involve measurement of the secondary structure (via circular dichroism), state of aggregation (via gel filtration), critical micelle concentration, phase transition temperature (for lipid bilayers), and so on.

Samples of proteins, membranes, or nucleic acids should have as little turbidity as possible. Microfiltration or centrifugation of samples is often necessary. Membrane systems are probably the most difficult to study, due to their unavoidable turbidity and tendency to settle. Because Rayleigh scattering from membrane samples is so intense, it is important to use a fluorophore that has a large Stokes shift. It goes without saying that the homogeneity of a sample is of paramount importance. Minor fluorescent impurities can make a significant contribution and can result in downward-curving Stern–Volmer plots, regardless of whether such impurities are quenched more or less easily by the quencher.

Other good fluorescence practices should be used. The absorbance of the sample at the excitation wavelength should normally be kept low (i.e., 0.20 or below) to ensure a good linear response of the signal. Usually, unpolarized excitation light is used. But such "natural" light will still lead to a photoselection of fluorophores that have their electronic oscillators in the vertical plane of the laboratory. Shinitzky (1972) has argued that this photoselection will lead to fluorescence intensity measurements (i.e., at right-angle detection) which are underestimates of the true intensity and that this can result in Stern–Volmer plots that deviate from linearity. This effect depends on the fluorescence anisotropy, r, of the sample. As the r value is increased by a dynamic quenching process, the measured fluorescence intensity (without polarizers) will become increasingly smaller than the true, total emission intensity. However, since the r values of fluorophores, in aqueous solution at room temperature, are usually less than 0.2, this effect is negligibly small in most cases. Only in the most rigorous studies, and when r values are high, should this effect be included. Experimentally this polarization effect can be avoided by using excitation or emission polarizers set at the "magic angles" (Badea and Brand, 1979).

4. DATA ANALYSIS

The fitting of the Stern–Volmer equation (6) to data can be trivial if the plot of F_0/F versus [Q] is linear. The quenching constant, K_{sv}, is just the slope of the line. If a Stern–Volmer plot curves upward or downward, a nonlinear least-squares (NLLSQ) analysis is needed. That is, Eq. (6) or (12) should be fitted to the data with an NLLSQ fitting procedure. Actually, it is preferable to fit the equations in the following form [for n equals 2, or possibly 3, in Eqs. (19) and (20); here $\exp(V[Q])$ is used for the static term]:

$$F/F_0 = \frac{1}{(1 + K_D[Q])\exp(V[Q])} \tag{18}$$

$$F/F_0 = \sum_{i=1}^{n} \frac{f_i}{1 + K_{sv,i}[Q]} \tag{19}$$

$$F/F_0 = \sum_{i=1}^{n} \frac{f_i}{(1 + K_{sv,i}[Q])\exp(V_i[Q])} \tag{20}$$

In these equations the experimental signal, F/F_0, is fitted directly, whereas in Eqs. (6) and (12), the inverse of the experimental signal appears. Fitting of the inverted data would give an improper weighting to the data [unless proper weighting factors are introduced (Dowd and Riggs, 1965)].

Computer programs for NLLSQ fitting of these or similar Stern–Volmer equations have been presented by Acuna *et al.* (1982) and Stryjewski and Wasylewski (1986). We have an NLLSQ program that we will make available upon request. Also, commercial programs, such as ASYST (MacMillan Software, New York, N.Y.) and ENZFITTER (Biosoft, Milltown, N.J.), have NLLSQ components that can be modified to fit these equations.

Nonlinear Stern–Volmer plots have often been analyzed via the following double reciprocal plot introduced by Lehrer (1971; Lehrer and Leavis, 1978) (here we neglect static quenching; $\Delta F = F_0 - F$).

$$\frac{F_0}{\Delta F} = \left(\sum_{i=1}^{n} \frac{f_i K_{sv,i}[Q]}{1 + K_{sv,i}[Q]} \right)^{-1} \tag{21}$$

If there are only two fluorescent components ($n = 2$) and if one component is completely inaccessible to the quencher (i.e., $K_{sv,2} = 0$), then Eq. (21) becomes

$$\frac{F_0}{\Delta F} = \frac{1}{f_1} + \frac{1}{f_1 K_{sv,1}[Q]} \tag{22}$$

The latter equation is a straight-line function; from the intercept and slope of a plot of $F_0/\Delta F$ versus $1/[Q]$, one can obtain f_1, the fraction of the fluorescence that is accessible to the quencher, and $K_{sv,1}$, the quenching constant for this accessible component. Figure 1C shows examples of such a plot.

It is important to stress that Eq. (22) is valid (and a plot $F_0/\Delta F$ versus $1/[Q]$ will be linear) only for the case in which there is a class of completely inaccessible fluorophores. If a plot of $F_0/\Delta F$ versus $1/[Q]$ is not linear, then forced analysis with Eq. (22) can be misleading. [It is better to use an NLLSQ analysis of Eq. (18)–(20) in such cases.] To illustrate the problems associated with analysis of curved plots of $F_0/\Delta F$ versus $1/[Q]$, Lehrer and Leavis (1978) presented simulations, to which the reader is referred. To excerpt a few examples of their simulations, if $K_{sv,1} = 10 \text{ M}^{-1}$, $K_{sv,2} = 2 \text{ M}^{-1}$, $f_1 = 0.5$, and $f_2 = 0.5$, a case in which one fluorophore is much more (5×) accessible to quencher than the other, forced analysis via Eq. (22) (i.e., taking slopes and intercepts of the data at low [Q]) will give an effective $K_{sv,1} = 8.67 \text{ M}^{-1}$ and effective $f_1 = 0.69$ for the accessible component. If $K_{sv,1} = 10 \text{ M}^{-1}$, $K_{sv,2} = 2$ M^{-1}, $f_1 = 0.25$, and $f_2 = 0.75$, graphical analysis via Eq. (22) yields an effective $K_{sv,1} = 7.0 \text{ M}^{-1}$ and effective $f_1 = 0.57$. Both of these examples show that forced analysis of data via Eq. (22), when the assumption that $K_{sv,2} = 0$ does not hold, leads to estimates of quenching parameters for the more accessible component that underestimate (for $K_{sv,1}$) and overestimate (for f_1) their true values.

Graphical presentation of data as plots of $F_o/\Delta F$ versus $1/[Q]$ has its merits, however, as it emphasizes differences in data at low $[Q]$. However, such plots can also mask the true extent of quenching that occurs. In Fig. 3A I present hypothetical plots of $F_o/\Delta F$ versus $1/[Q]$ that have an ordinate and abscissa span and slopes similar to published plots. To the untrained observer (ignoring the error bars for a moment) this may look like good data and one might proceed to analyze these data with Eq. (22) to obtain effective $K_{sv,1}$, and f_1, values. In fact, the range of $F_o/\Delta F$ points is such that only about 0.6–1.3% total quenching (i.e., total percent change in F_o) for the top data set and only 1.6–3.3% total quenching for the bottom data set has actually occurred. When you consider that most fluorescence intensity measurements can only be made with a precision of about ±0.25%, some of the error bars in the top set of points would span a vertical distance almost as large as the ordinate range shown! As illustrated in Fig. 3B, if one instead plots these fluorescence quenching data directly as F/F_o versus $[Q]$, a much more moderate interpretation would likely be given (i.e., that there is little quenching).

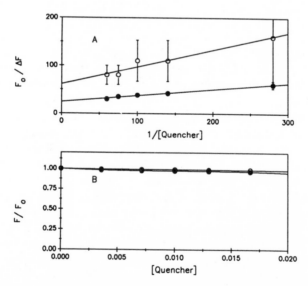

FIGURE 3. (A) Plot of $F_o/\Delta F$ versus $1/[Q]$ for mimicked data. See text. Error bars assume a reading error of ±0.25% for F values. (B) Replot of the above data as F/F_o versus $[Q]$; this direct plot shows that very little quenching was actually achieved. Note that forced analysis of the top data set with Eq. (22) will yield an apparent $f_1 \approx 0.016$ and an apparent $K_{sv,1} = 180\ M^{-1}$! Actually the effective K_{sv}, defined as $\sum f_i K_{sv,i}$, is only $\approx 3.0\ M^{-1}$ for the top data set.

The fitting of nonlinear Stern–Volmer plots, whether they are upward or downward curving, is difficult, even with NLLSQ programs. The plots may deviate only slightly from linearity and yet investigators may seek to recover anywhere from two to five fitting parameters (i.e., f_i, $K_{sv,i}$, V_i). The use of NLLSQ programs with proper weighting is the suggested approach (simultaneous analysis of multiple data sets is better; see below), but most fits are arbitrary and depend on which Stern–Volmer equation (i.e., with or without a static component, or for $n = 1, 2$, or 3) is selected. As a way to avoid the arbitrary nature of the fits, a useful approach is to compare the effective quenching constant, $K_{sv,eff}$. This $K_{sv,eff}$ is defined as the initial slope of the Stern–Volmer equation. $K_{sv,eff}$ is thus approximately equal to $\sum f_i(K_{sv,1} + V_i)$, that is, the weighted sum of the dynamic plus the static quenching constants for all components. If there is only one component, then $K_{sv,eff} = K_{sv} + V$. It is easy to determine this initial slope via NLLSQ or by plotting $(F_o/F - 1)/[Q]$ versus [Q] and extrapolating to [Q] $= 0$. Also, the inverse slope (*not* the intercept divided by the slope) of Eq. (21) is equal to $K_{sv,eff}$. While all quenching contributions are lumped together in this way, it is much easier to reproducibly recover $K_{sv,eff}$ values and comparative quenching studies can often be made more easily with this parameter. In some cases, however, it is desirable to separate $K_{sv,i}$ and V_i values and to determine f_i contributions.

Also, the need to maintain a constant ionic strength, when using a charged quencher, must be remembered. Recall in Fig. 2, for the KI quenching of the single Trp in staphylococcal nuclease, that, when the ionic strength was not held constant, a downward-curving Stern–Volmer plot was observed. If this electrostatic effect were not recognized, and these data were analyzed via Eq. (22) (assuming $K_{sv,2} = 0$), one might mistakenly conclude that nuclease has one class of Trp residues, comprising 18% of the total intensity (i.e., $f_1 = 0.185$) with an effective quenching constant, $K_{sv,1}$, of 13.5 M^{-1}, and another unquenchable class of Trp residues. When the salt concentration is maintained at 1.0 M, a linear Stern–Volmer plot is observed with $K_{sv} = 1.32$ M^{-1}.

The simultaneous analysis of a number of linked data sets is a procedure that has recently been popularized (Knutson *et al.*, 1983b; Beechem *et al.*, 1990) and that offers significant advantages. For example, solute quenching data obtained at more than one emission wavelength can be simultaneously analyzed via Eq. (19) or (20), using NLLSQ. Not only does this improve the confidence in recovered $K_{sv,i}$ values, but it also enables the wavelength dependence of the f_i to be evaluated. Wasylewski *et al.* (1988) performed acrylamide and iodide quenching on a number of proteins that possess two Trp residues. Figures 4–6 show how they were able to obtain individual $K_{sv,i}$ values for the Trp residues of the protein S. *aureus* metalloprotease and were also able to resolve the emission contours of the two Trp residues. Knutson

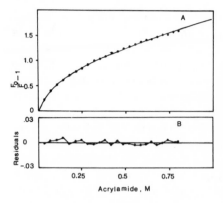

FIGURE 4. (A) Stern–Volmer plot for the acrylamide quenching of staphylococcal metalloprotease. Conditions: pH 9.0, 0.05 M Tris-HCl buffer with 0.01 M $CaCl_2$, 20°C, excitation at 297 nm, emission at 336 nm. Solid line is least-squares fit of Eq. (14) with $K_{sv,1}$ = 14.1 M^{-1}, $K_{sv,2}$ = 0.5 M^{-1}, and f_1 = 0.52. (B) Residual plot of the fit. Reprinted from Wasylewski *et al.* (1988) with permission from the *European Journal of Biochemistry*.

et al., (1983b) have used a "global analysis" procedure to analyze quenching data obtained at multiple emission wavelengths for model compounds and proteins. Our group has used this procedure for phosphoribulokinase (Ghiron *et al.*, 1988).

All of the above data analysis procedures make the assumption that the dynamic quenching of an individual fluorophore will follow the Stern–Volmer equation (9). That is, the assumption is made that individual Stern–Volmer plots are linear [or that Eq. (7) is followed if static quenching occurs]. There are reasons why such individual Stern–Volmer plots will be downward curving. Fluorescence lifetime studies show that individual Trp residues in proteins can be multiexponential (Beechem and Brand, 1985). This means that there may be multiple K_{sv} values for an individual Trp residue, even if there is only a single k_q (see Eftink, 1989, for further discussion). Also, with charged quenchers, the electrostatic effects discussed in Section 5.3 will

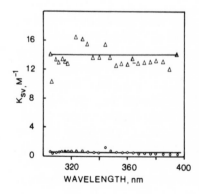

FIGURE 5. Plots of Stern–Volmer constants for the acrylamide quenching of staphylococcal metalloprotease as a function of emission wavelength. Conditions given in Fig. 4. An average $K_{sv,1}$ = 14.13 M^{-1} is found for the more accessible Trp residue and $K_{sv,2}$ = 0.513 M^{-1} for the less accessible residue. Reprinted from Wasylewski *et al.* (1988) with permission from the *European Journal of Biochemistry*.

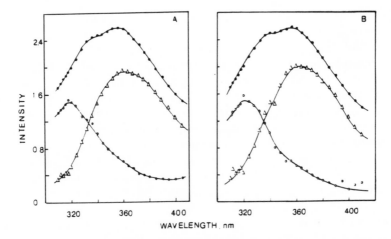

FIGURE 6. Resolution of the fluorescence subspectra of staphylococcal metalloprotease. The upper spectra (●) are the total fluorescence spectra of the protein; ○, subspectra for the less accessible Trp residue; △, subspectra for the more accessible residue. (A) Resolution with $K_{sv,1}$ and $K_{sv,2}$ fixed at their average values of 14.13 M^{-1} and 0.513 M^{-1}, respectively. (B) Resolution with $K_{sv,1}$ and $K_{sv,2}$ allowed to float at each wavelength. Reprinted from Wasylewski *et al.* (1988) with permission from the *European Journal of Biochemistry.*

result in a downward-curving Stern–Volmer plot, particularly if a high and constant ionic strength is not employed.

5. EXPERIMENTAL STRATEGIES

Most solute quenching experiments are designed to determine the accessibility, to a solute quencher, of an intrinsic or extrinsic fluorophore that is attached to a biomacromolecule. By variation of other conditions, such as temperature, pH, ionic strength, viscosity, or by altering the chemical nature of the fluorophore, quencher, or biomacromolecule, an investigator may gain further insights about the topography of the fluorophore–biomacromolecule. In this section I will discuss the strategies of varying these conditions and reactants.

5.1. Fluorescence Lifetime Measurements

First, however, it should be reiterated that the accessibility to quencher, in terms of quenching constant values, is the parameter that is determined in steady-state quenching experiments. A more useful parameter is the quenching rate constant, k_q, for a quencher–fluorophore reaction. Since the dynamic quenching constant, K_{sv}, is equal to $k_q\tau_o$, it is necessary to know the value of the unquenched fluorescence lifetime, τ_o, in order to calculate k_q. This requires a time domain or frequency domain fluorometer, the availability of which is constantly increasing. If a quenching reaction can be described by a single K_{sv} and if the lifetime is a monoexponential, then there is no problem in calculating k_q. However, if the fluorescence decay is a double- or multi-exponential and/or the solute quenching profile must be described by more than one $K_{sv,i}$, then the interpretation can be difficult.

Some of the more detailed quenching studies in recent years have involved the measurement of fluorescence decay profiles (time or frequency domain) as a function of quencher concentration (Ross et al., 1981; Torgerson, 1984; Robbins et al., 1985; Demmer et al., 1987; Chen et al., 1987; Eftink and Wasylewski, 1989). For example, Ross et al. (1981) measured the fluorescence decay of the two-Trp protein, horse liver alcohol dehydrogenase (LADH), at various concentrations of KI. They found the fluorescence decay to be a double-exponential and further found that one of the lifetimes of 3.4 nsec was unchanged by the addition of KI, whereas the other lifetime of 7.8 nsec was quenched by KI in a Stern–Volmer manner [i.e., Eq. (9)] with a $k_q = 0.76 \times 10^9$ M^{-1} sec^{-1}. In conjunction with other information about LADH, the 3.4-nsec, unquenchable component was assigned as Trp-314, and the 7.8-nsec, quenchable component was assigned as Trp-15. More recently, Demmer et al. (1987) have reinvestigated the iodide quenching of LADH with time-domain fluorescence methods. Their results are in general agreement with Ross et al., but Demmer et al. also found that the decay of Trp-15 is best fitted as a double exponential. In Fig. 7 are shown the resulting Stern–Volmer plots for the two decay times for Trp-15; note that the slope ($=k_q$) is the same for each decay time, as would be expected.

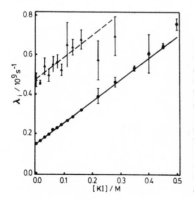

FIGURE 7. Recovered values of the two fluorescence decay times $(1/\lambda_i)$ for Trp-15 of horse liver alcohol dehydrogenase as a function of added KI. ●, the long-lived component (about 6.8 nsec) of this Trp residue; ▲, the short-lived component (about 2.1 nsec) of this residue. The slope of each λ_i versus [Q] plot is the same, within experimental certainty, and gives $k_q \approx 1 \times 10^9$ M^{-1} sec^{-1}. The total decay of the protein includes a 4.2-nsec component, assigned to Trp-314, which is not quenched by KI. Reprinted from Demmer et al. (1987) with permission from *Photochemistry and Photobiology*.

FIGURE 8. Multifrequency phase (open symbols) and modulation (closed symbols) data for cod parvalbumin in the absence (O, ●) and presence of the following concentrations of acrylamide: (△, ▲), 0.46 M; (□, ■), 1.06 M. These data, along with three other data sets at other [Q], were simultaneously fitted to a model in which the fluorescence decay of the protein is described as a Lorentzian distribution of decay times centered at 3.4 nsec and having a full-width-half-maximum of 1.7 nsec and with a single quenching rate constant of 0.18×10^{9} M^{-1} sec^{-1}. Reprinted from Eftink and Wasylewski (1989) with permission from *Biochemistry*.

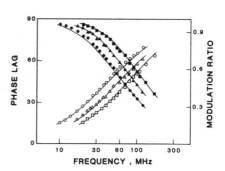

Our lab has performed similar lifetime solute quenching studies with cod parvalbumin using the frequency domain method (Eftink and Wasylewski, 1988a). Shown in Fig. 8 are phase/modulation profiles for this protein in the presence of different acrylamide concentrations. For this system the fluorescence decay was fitted as a Lorentzian distribution of decay times and the data in Fig. 8 (plus three other sets of curves at addition [Q]) were described with only three fitting parameters, the quenching rate constant, k_q, the center of the lifetime distribution, $\bar{\tau}$, and the full-width-at-half-maximum of the distribution, Δ.

Fluorescence lifetime/solute quenching data contain more information than steady-state/solute quenching data, but, in fact, the complexity of the former can be overwhelming when the decays are complex. Even in such cases, however, steady-state/solute quenching studies can usually still be interpreted in a qualitative, yet useful, way. If the decay of a fluorophore is multiexponential, it is still possible to describe this decay with an average lifetime ($= \sum f_i \tau_i$) or in terms of the mean lifetime of a distribution. Such average lifetimes, together with $K_{sv,eff}$ values from steady-state quenching, can give average k_q values. By using some of the strategies below, the variations in k_q, as conditions are altered, can yield molecular insights.

If a researcher does not have fluorescence lifetime instrumentation, he or she can still determine a quenching constant for a fluorophore–biomacromolecule relative to that for a model compound. This ratio will generally reflect the ratio of the k_q values for the fluorophore in the macromolecule system to that of the free state. An even better approximation of this k_q ratio is possible if the K_{sv} ratio is divided by the ratio of fluorescence quantum yields of fluorophore in the attached and free states.

The k_q value for a fluorophore attached to a biomacromolecule (i.e., a Trp residue on a protein) should be lower than that for the free fluorophore.

This is because the translational and rotational diffusion coefficient of the bound fluorophore is smaller than the value for the free fluorophore (Johnson and Yguerabide, 1985). This reduction in k_q by attachment is separate from any shielding of the fluorophore by the biomacromolecule, which will further reduce k_q. Johnson and Yguerabide (1985) have shown that the maximum k_q values for an unshielded, attached fluorophore will depend on the size of the biomacromolecular substrate, but in general one can expect the maximum k_q for an attached fluorophore to be about half that for a free fluorophore.

5.2. Variation of Quencher Size, Charge, and Polarity

It is easy to imagine performing a comparative study of the quenching of a Trp residue in a protein by a series of quenchers of differing size. As with solvent perturbation studies, this could reveal topographical features of the protein surface near the Trp residues (i.e., the size of a crevice leading to a Trp residue). While such experiments can be imagined, in practice it is difficult to select a group of quenchers that have the same quenching mechanism but differ just in size.

Our lab has compared the quenching of protein Trp residues by acrylamide and the slightly larger succinimide (Eftink and Ghiron, 1984); both of these quenchers are neutral and polar and probably quench by an electron transfer mechanism. We found, however, that the efficiency of quenching by succinimide is lower than that by acrylamide and that this results in a significant dependence of succinimide quenching reactions on the microenvironment of indole. Useful insights were possible by comparing succinimide and acrylamide, but differences between these two probably reflect the differences in their quenching efficiency.

Due to its small size, molecular oxygen is a very good quencher of the fluorescence of Trp residues in proteins (Lakowicz and Weber, 1973b). A comparison of the oxygen k_q with that for other, larger quenching probes is a good strategy (Calhoun et al., 1983). In fact, we have shown (Eftink, 1991) that there is a good correlation between the oxygen k_q and the acrylamide k_q for the quenching of single-Trp proteins. Those Trp residues that have the smallest k_q for the larger, polar acrylamide are also the residues that have the smallest k_q for oxygen (i.e., the single Trp in apoazurin from *Pseudomonas aeruginosa* has relatively low k_q for both quenchers). Of course, the k_q for oxygen are all larger than those for acrylamide, for the series of proteins, but the good correlation between the data sets indicates that (1) the same kinetic principles govern the action of each type of quencher and (2) the larger quencher experiences more steric resistance, as would be expected.

Variation in the charge and/or polarity of the quencher is a routine and effective strategy. TCE is an apolar quencher. With certain proteins, TCE appears to accumulate in oily regions and to cause the protein to expand and unfold. At low concentrations, TCE thus may preferentially quench fluorophores in oily regions, as opposed to acrylamide and charged quenchers, which should prefer to quench fluorophores near the surface of the biomacromolecule. This pattern should particularly be true for fluorophores embedded in bilayer membranes.

Iodide is a dependable anionic quencher, but the choice of a corresponding cationic quencher is not as clear. Thallous ion, Tl^+, may prove to be the best cationic quencher, if it does not cause precipitation of the biomacromolecular system. Charged quenchers have the greatest selectivity for solvent-exposed fluorophores and they also are influenced by the presence of charged groups near the fluorophores (see below). Ionic quenchers, such as iodide and N-methyl-picolinium, have been particularly useful in studies with membrane systems, since these charged quenchers are very selective for fluorophores (i.e., surface Trp residues of proteins) on the surface of the membrane (Shinitzky and Rivnay, 1977; Kauffman *et al.,* 1983; Pownall and Smith, 1974).

5.3. Variation of Ionic Strength and pH

Charged quenchers can be used to determine the sign of the electric potential near a fluorophore, through studies in which the ionic strength is varied. This strategy has been used by several researchers and was presented in a rigorous manner by Ando *et al.* (1980). The readers are referred to their works for equations describing the dependence of quenching rate constants on ionic strength, μ. In Fig. 9 is shown an example of the type of information that can be obtained. Three quenchers of different charge, iodide ion, acrylamide, and thallous ion, were used to quench the fluorescence of the fluorophore, N-(p-(2-benzimadazolyl)phenyl)maleimide (BIPM), which was covalently attached to heavy meromyosin. The k_q for acrylamide was found to have little dependence on μ. For iodide and thallous ion, however, increasing μ caused their k_q values to increase and decrease, respectively. This indicates that there is a negative electric potential in the neighborhood of BIPM. In other studies, Ando *et al.* (1980) attached BIPM to a second site on the protein. This site was found to have almost no nearby electric potential. Binding of ATP, however, resulted in the introduction of negative potential near the second site.

In proteins, the electric potential at sites on a protein is primarily due to the charges on amino acid side chains. As these will undergo protonic equilib-

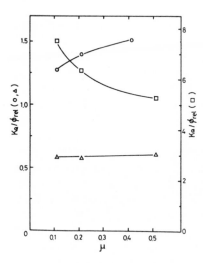

FIGURE 9. The ionic strength dependence of the relative rate constants (K_Q/Φ_{rel}) of quenching of heavy meromyosin labeled with the fluorescent probe, BIPM, at a particular sulfhydryl group denoted as SH_1. $\triangle$, acrylamide as quencher; $\bigcirc$, iodide; $\square$, Tl^+. Reprinted from Ando *et al.* (1980) with permission from the *Journal of Biochemistry*.

ria, the electric potential can be varied by changing pH. Thus, by studying the pH dependence of the k_q for quenching by a charged quencher, one should be able to determine the charge and pK_a of an ionizing side chain near a fluorophore. A simple way to do the latter is to perform a pH titration of the fluorescence of a sample in the absence and presence of an ionic quencher, such as KI (Lehrer, 1971; DeWolf *et al.*, 1987). Of course, these variations in pH or μ may cause their effect indirectly by promoting a change in the conformation of the biomacromolecule. A companion study of the quenching by a neutral quencher should show if this is the case.

5.4. Variation of Temperature, Pressure, and Viscosity

These variables have been exploited by our group and others to characterize the nature of the quenching mechanism. With internal Trp residues in proteins there is a question as to whether neutral quenchers, such as acrylamide, quench by a mechanism that involves penetration of the quencher into the protein or segmental unfolding of the protein to reveal the internal Trp to the solvent. I have discussed these matters in a separate review on fluorescence quenching (Eftink, 1991) and will not give a lengthy rehash of the stories here. Studies in which temperature, pressure, and bulk viscosity are varied have provided useful insights regarding this question.

Determination of the solute quenching k_q as a function of temperature enables an apparent Arrhenius activation energy, E_a, to be obtained. If the rate-limiting step for the quenching reaction is diffusion of the quencher through the bulk solvent, then one expects the E_a to be about 4 kcal/mole (the E_a for the inverse of the viscosity of water). For certain internal Trp residues in proteins, acrylamide quenching E_a values have been found that are much smaller or larger than 4 kcal/mole (Eftink and Ghiron, 1978). This indicates that some type of fluctuation in the protein contributes to the rate-limiting step.

Likewise, from pressure dependence studies one can obtain an apparent activation volume, $\Delta V^{\ddagger}$, for a quenching reaction. We have determined some $\Delta V^{\ddagger}$ for the acrylamide quenching of Trp residues in proteins (Eftink and Wasylewski, 1988b). In general, the $\Delta V^{\ddagger}$ are very small and are consistent with the involvement of small-amplitude fluctuations in the protein. Such pressure studies require a special high-pressure cell (1 to about 4000 atm) and are best done with fluorescence lifetime measurements.

Variation in temperature and pressure may also cause changes in k_q through an induced phase change or melting of the biomacromolecular assembly. For example, Gomez-Fernandez *et al.* (1985) have studied the temperature dependence of the acrylamide quenching of the Trp fluorescence of Ca^{2+}, Mg^{2+}-ATPase, reconstituted with either dimyristoylphosphatidylcholine or dipalmitoylphosphatidylcholine. Breaks in plots of $K_{sv,eff}$ versus temperature were seen; these breaks correspond to the phase transition temperature of the phospholipids. Larger $K_{sv,eff}$ were seen for the liquid crystalline phase of the phospholipid matrix, indicating a linkage between membrane fluidity and the accessibility of Trp residues to the quencher.

Since fluorescence quenching reactions are often diffusion-limited, these reactions can be attenuated by increasing the viscosity, η, of the bulk (i.e., by adding glycerol or sucrose). With fluorophores attached to biomacromolecules, a quenching k_q is expected to be inversely proportional to η if the fluorophore is on the surface. If the fluorophore is within the biomacromolecular structure (i.e., internal Trp in protein), the quenching k_q may depend less on viscosity (Eftink and Hagaman, 1987). Of course, an absence of an η dependence can be taken as an indication that a quenching reaction is a static process. Fluorescence lifetime quenching measurements are needed to confirm whether a quenching process is static or dynamic.

The variation of viscosity is easy to achieve, but viscous solutions present some problems. Viscous solutions are difficult to mix. The solubility of oxygen in glycerol–water mixtures is low, thus making oxygen quenching studies difficult. Absorptive screening by acrylamide is a greater problem in glycerol–water mixtures. Also, the purity of viscogens, such as glycerol, sucrose, or propylene glycol, is important. With the high concentrations used, fluorescent impurities can be a problem.

The viscosity dependence of the k_q for the acrylamide or oxygen quenching of free indole (and presumably other small fluorophores) does not follow the Stokes–Einstein relationship in glycerol–water or sucrose–water mixtures. The degree of departure is typical of other diffusion-limited reactions, however (Alwatter *et al.*, 1973); the departure simply reflects the inadequacy of the Stokes–Einstein theory to predict the diffusional behavior of small molecules. We have presented the η dependence of acrylamide and oxygen k_q values, for the quenching of indole, to serve as reference values for other η dependence studies with these quenchers (Eftink and Ghiron, 1987a; Eftink and Hagaman, 1987).

5.5. Variation of the State of the Biomacromolecule

Changes in the accessibility of a fluorophore can report a conformational change in a biomacromolecule, which may be induced in may ways. In the following section we will discuss specific examples of the use of solute quenching to study such conformational changes. Experiments are often designed to induce such changes or to compare different forms of a biomacromolecule. In general, alterations may include (1) conformational changes or aggregation/disaggregation induced by temperature, ionic strength, pH, denaturants, and so on and (2) conformational changes induced by the binding of specific ligands (see Eftink, 1991, for a list of example studies). Or comparison may be made of (3) homologous biomacromolecules (Edwards and Silva, 1986; Sommers and Kronman, 1980), including proteins made by site-directed mutagenesis (Hansen *et al.*, 1987).

5.6. Variation of the Fluorophore

Most applications of solute fluorescence quenching involve changes in the quencher, the biomacromolecule's state, or reaction conditions, but in some cases it can be useful to change the fluorophore attached to a biomacromolecular assembly. This is true if the mode of attachment may differ from different fluorophores and if such differences wish to be revealed. An example in which the biomacromolecule and the quencher remained the same, but the bound fluorophore was changed, is a study by Zinger and Geacintov (1988) of the quenching of different fluorescent drugs that were bound to double-stranded DNA. The purpose of this study was to characterize the different binding modes, i.e., whether the different drugs bind to DNA by intercalating, by binding in the minor groove, and so on. Another example is the study of a positional series of *n*-(9-anthroyloxy)-labeled fatty acids. By incorporating these into lipid bilayers and quenching by ionic (I^- and Cu^{2+}) and neutral (acrylamide) quenchers, Chalpin and Kleinfeld (1983) were able to characterize the depth of penetration of the *n*-(9-anthroyloxy) groups on the various fatty acids.

5.7. Use of Quencher-Lipid Molecules

Brominated or nitroxide-labeled fatty acids and phospholipids are a very useful set of probes for the depth of penetrations, into a lipid bilayer or vesicle, of proteins or other fluorescing species. In bilayers, these quencher-lipids, many of which are commercially available, will align with their alkyl chains perpendicular to the bilayer plane. The quencher group can be positioned at different places along the chain. Thus, it is possible to determine whether the Trp of an embedded protein, for example, are located near the bilayer's center, by use of different quencher-lipid probes.

Since the fatty acids or phospholipids, to which the quencher moiety is attached, will undergo rather slow (on the nanosecond time scale) lateral motion, the quenching by these quencher-lipids is considered to be largely a static process. In employing and analyzing data with these quencher-lipids, two strategies can be used. One is to measure the fluorescence of the sample as a function of increasing mole fraction, X_Q, of the quencher-lipid in the phospholipid bilayer. With the following equation, derived by London and Feigenson (1981), this approach can yield an estimate of the fluorophore's position in terms of the maximum number, n, of quencher-lipids that can be adjacent to the fluorophore:

$$\frac{F - F_{min}}{F_0 - F_{min}} = (1 - X_Q)^n \tag{23}$$

Here F_{min} is the minimum fluorescence when the sample is fully quenched. Plots of $(F - F_{min})/(F_0 - F_{min})$ versus X_Q can be fitted for various values of n. If n is found to be as large as 6, this indicates that the fluorophore is located within the bilayer and is not shielded (i.e., not buried within a protein's structure) from lipid molecules.

A second strategy is to compare the quenching by two or more different quencher-lipids having their quenching moiety at different positions along an alkyl chain (i.e., nitroxide-labeled fatty acids at positions 5, 10, or 12). For such a study, Chattopadhyay and London (1987) have presented equations to be used to determine the depth of bilayer penetration of a fluorophore (see next section).

6. EXAMPLES OF TOPOGRAPHICAL SOLUTE QUENCHING STUDIES

In this section I will briefly discuss selected examples of the use of solute quenching to obtain topographical information about biomacromolecules and their assemblies. The emphasis here will be to cite examples that employ some of the strategies mentioned above and that illustrate the usefulness of solute quenching in studying a variety of important biochemical systems.

Some of the early solute quenching studies showed that the quenching patterns for denatured proteins are significantly different from the folded, globular structures (Lehrer, 1971; Eftink and Ghiron, 1976b). Further exploitation of this is evident in two recent works involving the conformational state of proteins. Havel *et al.* (1988) have studied the unfolding and aggregation of the single-Trp protein, bovine growth hormone ($M_r \sim 22,000$). The equilibrium unfolding of this globular protein, induced by guanidine-HCl, can be represented as

$$N \rightleftharpoons I \rightleftharpoons U$$
$$\updownarrow$$
$$I_n$$

where N and U are the native and completely unfolded states, I is a monomeric intermediate, and I_n is an aggregated intermediate (for which n appears to be between 3 and 5). Havel *et al.* have studied the iodide, acrylamide, and TCE quenching of the Trp residue of each of these conformational states. Their results are summarized in Table III as effective k_q values (obtained by separate measurements of the average fluorescence lifetimes). They found the Trp residue to become more exposed to all three quenchers as the conformation was changed from N to I to U. In the associated intermediate state, I_n, the Trp was found to have reduced exposure, especially to iodide. It was suggested that the low k_q for iodide is due to an electrostatic shielding of the Trp in the aggregated state.

TABLE III. Parameters for the Quenching of the Fluorescence of Human Growth Hormone by Iodide, Acrylamide, and Trichloroethanol[a,b]

Conformational state	k_q		
	Iodide	Acrylamide	Trichloroethanol
N	0.80	1.1	0.75
I	0.7	1.4	1.0
I_n	0.19	0.8	0.75
U	0.72	2.4	2.5

[a] See Havel *et al.* (1988) for conditions. The units of k_q are $\times 10^9$ M^{-1} sec^{-1}. These k_q values are calculated using average lifetimes. For trichloroethanol, significant V values (static quenching) were reported for the I and I_n states.
[b] Summarized from Havel *et al.* (1988) with permission of the authors.

The denaturation of the tetrameric protein, pigeon liver malic enzyme, has similarly been studied by Lee *et al.* (1988). With the addition of quanidine-HCl, this enzyme appears to undergo a stepwise tetramer → monomer → random coil transformation. Using acrylamide and iodide as quenchers of Trp (5 per subunit) fluorescence, the biphasic dissociation/unfolding was observed as an increase in quenchability.

The use of solute quenching to monitor ligand-induced changes in the conformation of a protein can be illustrated by dozens of examples. Here only a couple will be cited. The binding of specific inhibitors and transition-state analogues to the enzyme human adenosine deaminase was studied via solute quenching by Phillips *et al.* (1987). Acrylamide, TCE, and iodide were used to determine the accessibility of the Trp fluorescence of this protein. The apolar TCE was found to be the most effective quencher, suggesting that the Trp residues of this protein are located in a relatively hydrophobic environment. The interaction of this protein with four specific ligands was studied and characteristic changes in the solute quenching patterns were seen for each.

I again mention the work of Ando *et al.* (1980) with heavy meromyosin (HMM). The addition of the substrate ATP to HMM results in slow steady-state hydrolysis. Ando *et al.* studied the solute quenching, of the fluorescence of extrinsic fluorophores attached to HMM, during this hydrolysis reaction (i.e., studies of an enzyme–ADP reaction intermediate produced at steady state). They used neutral, anionic, and cationic quenchers and also varied ionic strength (see Fig. 9). They were able to sense changes in the accessibility of the fluorophore and also in the charge density around the fluorophore produced by adding the substrate.

Most successful solute quenching studies have been done with biomacromolecules that have only one or a few fluorophores, but useful information can also be obtained with much larger, more complicated systems. As an example, Hill *et al.,* (1986) have studied the quenching, by acrylamide and iodide, of the fluorescence of the ~ 50 Trp residues of cytochrome c oxidase, solubilized by lauryl maltoside. As shown in Fig. 10, a careful study of the iodide quenching showed little drop in fluorescence from 0 to 0.1 M KI; above 0.2 M KI, however, an abrupt increase in the degree of quenching was observed, indicating that the addition of KI induces a change in the structure of this protein.

Another large protein system studied by solute quenching is the Na^+, K^+-ATPase. Tyson and Steinberg (1987) used acrylamide to quench the ~ 16 Trp residues of this protein and observed changes in the overall Trp accessibilities as Na^+, ADP, K^+, and ouabain were added to the protein. As shown in Fig. 11 the binding of Na^+ and ADP seems to stabilize a common structure of the protein (E_1 form), as judged by the accessibility of Trp resi-

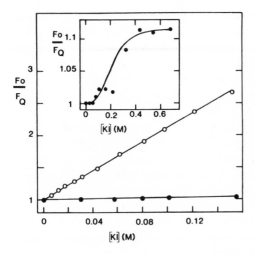

FIGURE 10. Stern–Volmer plot for the iodide quenching of cytochrome c oxidase (●) and N-acetyl-L-tryptophanamide (O). Conditions for the protein: 0.02 M Tris, pH 7.8, with 0l1 nN EDTA and 1 mg/ml lauryl maltoside, 25°C, excitation at 280 nm, emission at 328 mm. Ionic strength was maintained with NaCl. Little iodide quenching of the protein is seen below 0.15 M. Inset: iodide quenching at higher quencher concentrations. Reprinted from Hill *et al.* (1986) with permission from *Biochemistry.*

dues to acrylamide (i.e., similar modified Stern–Volmer plots). The binding of K^+ and ouabain, on the other hand, appears to stabilize another structure of the protein (E_2), as judged by the similar patterns in Fig. 12.

A couple of recent applications of solute quenching involve the selective incorporation, via solid-state chemical synthesis or site-directed mutagene-

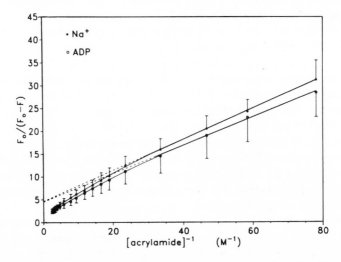

FIGURE 11. Plots of $F_0/\Delta F$ versus $1/[Q]$ for the acrylamide quenching of the fluorescence of Na^+,K^+-ATPase. Conditions: 0.1 M Tris-HCl, 0.1 mM EDTA, pH 7.5, 25°C, excitation at 295 nm. Data obtained in the presence of 3 mM Na^+ (●) and 30 μM ADP (O). Reprinted from Tyson and Steinberg (1987) with permission from the *Journal of Biological Chemistry.*

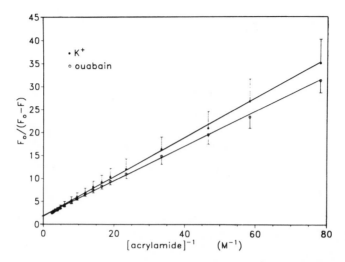

FIGURE 12. Plots of $F_0/\triangle F$ versus $1/[Q]$ for the acrylamide quenching of Na^+,K^+-ATPase in the presence of 0.5 mM K^+ (●) and 0.1 mM ouabain (○). Reprinted from Tyson and Steinberg (1987) with permission from the *Journal of Biological Chemistry*.

sis, of Trp residues in specific positions in a peptide or protein. These are, in my opinion, examples of the exciting new studies that are made possible by modern, protein engineering methods. O'Neil *et al.* (1987) have chemically synthesized a series of peptides having a single Trp residue at each of 16 positions along a 17-amino-acid chain. The particular peptide is capable of forming an α-helix and is an analogue of the myosin light chain kinase (MLCK) peptide, which interacts with calmodulin. Complexes between the MLCK peptide analogues and calmodulin (which has no Trp) were formed and the accessibility of the Trp residue to acrylamide quenching, and other fluorescence parameters, were determined as a function of the Trp position. Figure 13 shows the value for the acrylamide quenching constant (panel B), together with the fluorescence λ_{max} and the steady-state anisotropy, for the series of peptides. A periodic pattern is seen for each parameter. O'Neil *et al.* have interpreted this as indicating that the MLCK peptides are in a helical state when they are bound to calmodulin. Figure 14 illustrates a proposed mode of binding that explains the periodic fluorescence pattern.

The great potential of the solute quenching together with site-directed mutagenesis is illustrated by the elegant study of Hansen *et al.* (1987) with *E. coli* tet repressor and its mutants. This protein normally has two Trp residues at positions 43 and 75. Mutants were prepared that have a single Trp residue in either of the above positions (with the other position replaced with a phenylalanine). Iodide quenching studies with the two mutants and the wild

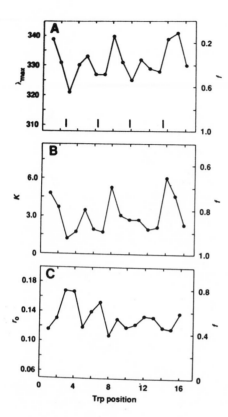

FIGURE 13. (A) Fluorescence λ_{max} of MLCK peptide: calmodulin complexes as a function of Trp position. (B) Acrylamide quenching K_{sv} of the complexes as a function of Trp position. (C) Anisotropy (λ_{ex} = 300 nm) of the complexes as a function of Trp position. See original article for definition of f. Reprinted from O'Neil et al. (1987) with permission from *Science*.

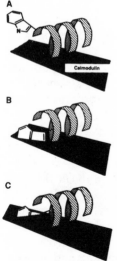

FIGURE 14. Schematic representation of the interaction of the α-helical MLCK peptide analogues with calmodulin and the different orientations that may be assumed by the single Trp residue as its position is changed from the N-terminus to the C-terminus. For structures in which the Trp residue is on the hydrophobic face of the helix and is closely associated with calmodulin (panel C), the largest changes in fluorescence properties are expected (i.e., blue λ_{max}, low K_{sv}, high anisotropy, r_o). Reprinted from O'Neil et al. (1987) with permission from *Science*.

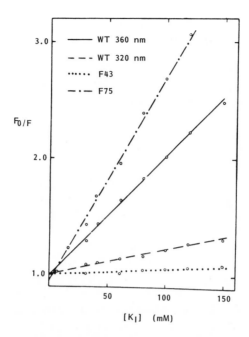

FIGURE 15. Stern–Volmer plots for the iodide quenching of *E. coli* tet repressor (wild type, WT) and its mutants (F43 and F75). Conditions: 5 mM Tris-HCl, pH 8.3, 0.1 mM EDTA, 5 mM MgCl$_2$, 0.1 mM dithiothreitol, 20°C, excitation at 280 nm, emission wavelength as indicated for WT, 328 nm for F43, and 352 nm for F75. Ionic strength maintained at 0.2 M with KCl. Reprinted from Hansen *et al.* (1987) with permission from the *Journal of Biological Chemistry.*

type are shown in Fig. 15. The mutant F75 (which has Trp-43) was found to be easily quenched by iodide; thus, Trp-43 is accessible to this charged quencher. The mutant F43 (which has Trp-75) was found to be quenched very weakly; thus, Trp-75 is relatively inaccessible to iodide. The quenching pattern for the wild type was found to be a composite of that for the individual Trp residues. Once the fluorescence contributions from the two Trp residues were separated and characterized, these residues were used as intrinsic reporters to study the binding of the inducer, tetracycline.

With membrane systems there are the possibilities of (1) using charged or polar quenchers to determine the solvent accessibility of fluorophores, (2) using apolar quenchers, which freely partition into and diffuse through the lipid domain, to determine the lipid coverage of a fluorophore and/or the rates of lateral diffusion of the quencher or fluorophore, or (3) using an aligned quencher-lipid (see Section 5.7), which has a quenching moiety positioned at various places along a lipid substrate, to quantify the depth of penetration of a fluorophore within a lipid bilayer. Examples of the first possibility are the study of Batenburg *et al.* (1985) of the use of iodide and acrylamide quenching to characterize the surface accessibility of cardiotoxin II of *Naja mossambica mossambica* (single Trp) associated with vesicles of cardiolipin, and the study of Georghiou *et al.* (1982) of the acrylamide quenching of bee venom melittin (single Trp) associated with vesicles of phosphatidylcholine.

For quenchers that partition into lipid domains, Mantulin *et al.* (1986) have used oxygen to quench the Trp fluorescence of complexes of human apolipoprotein A-I with dimyristoylphosphatidylcholine. From temperature and bulk viscosity dependence studies they found that the oxygen quenching rate constant primarily reflects the partitioning of this quencher between the bulk aqueous and lipid phases. The lateral diffusion coefficient of the membrane-incorporated quencher, ubiquinone, was determined in a study by Fato *et al.* (1986) of the ability of this quencher to quench the fluorescence of 12-(9-anthroyloxy)-stearic acid in phospholipid and mitochondrial membrane vesicles. This study first involved the evaluation of the partition coefficient of ubiquinone into the lipid component, by studying the quenching at different lipid/quencher ratios (see Section 2.3). Once this partition coefficient was determined (and hence the concentration of ubiquinone in the bilayer), the quenching rate constant was calculated. This rate constant was then related to the lateral diffusion coefficient of the quencher by the three-dimensional Smoluchowski equation [Eq. (10)] or a two-dimensional counterpart (Blackwell *et al.,* 1987).

A direct way to obtain topographical information about membrane systems is by the use of quencher-lipids that contain either bromo or nitroxide groups at different positions along a lipid molecule. Markello *et al.* (1985) prepared such a series of dibrominated phospholipids and used these to study the depth of penetration of cytochrome b_5 into lipid vesicles. They used both fluorescence intensity and lifetime measurements. The phospholipid dibrominated at position 6,7 was found to be the most effective quencher (as compared to quencher-lipids dibrominated at positions 9,10, 11,12, 15,16) and the quenching was found to be, at least partially, a dynamic process. There are probably three fluorescent Trp residues in cytochrome b_5 and these quenching results suggest that some of these residues are located at a depth of about 7 Å from the bilayer surface. Using Eq. (23) and the analysis procedure of London and Feigenson (1981), Markello *et al.* determined a value of $n = 3$ for the 6,7-dibromo phospholipid. This n value suggests that the quenched Trp residues can be surrounded by three phospholipid molecules.

Chattopadhyay and London (1987) have employed a set of nitroxide spin-labeled phospholipids (at positions 5, 10, and 12) to quench membrane-associated fluorescence probes. The probes contained the fluorophore, 7-nitro-2,1,3-benzoxadiazol-4yl (NBD), at various positions along a fatty acid tail (6 NBD PC and 12 NBD PC) or the head group (NBD PE) of a phospholipid or at the nonpolar end of cholesterol (NBD cholesterol). Shown in Fig. 16 are the structures and possible locations of the fluorophores. The results, in Fig. 17, show that the position 5 spin-labeled phosphatidylcholine (5 SL PC) is the most effective quencher (and 12 SL PC is the least effective quencher) for NBD PE, 6 NBD PC, and 12 NBD PC, indicating that the fluorophore lies close to the bilayer's surface in these cases. For NBD cholesterol, 12 SL PC is the best and 5 SL PC is the worst quencher; this indicates that the fluorophore is buried near the bilayer's center in this case. Chattopadhyay and London developed equations to enable a quantitation of the depth of burial of fluorophores by comparison of the quenching by two or more spin-labeled quencher-lipids.

Most applications of solute quenching to nucleic acid systems have focused on either the fluorescence of bound dyes (Zinger and Geacintov, 1988) or the Trp fluorescence of proteins that interact with nucleic acids. As an example of the latter systems, Kan *et al.* (1986) have studied the acrylamide quenching of the coat protein (two Trp) of alfalfa mosaic virus. They studied the protein alone (a dimer) and as a viroid particle obtained by the addition of RNA. The quenching rate constant was found to be relatively small (0.45×10^9 M^{-1} sec^{-1}) for the dimer and to decrease only moderately (to 0.3×10^9 M^{-1} sec^{-1}) upon assembling with the RNA to form the viroid particle. This suggests that the Trp residues are found neither at the dimer–dimer interface nor at the dimer–RNA interface.

In a final example, Ferguson and Yang (1986) have attached fluorescent probes, at different positions, to tRNA[fmet] and have observed changes in the accessibility of these probes to quenchers as the modified tRNA[fmet] binds to methionyl-tRNA synthetase.

7. CONCLUSION

Solute quenching reactions will probably continue to be a very useful tool for obtaining topographical information for biomacromolecular structures. Advances in data acquisition and analysis procedures and advances in the ability to incorporate (i.e., site-directed mutagenesis) fluorescent probes into biomacromolecules should enable more, productive studies in the future. Because the experiments are so easy to do, many researchers have used or will use this technique to study their system of interest. I hope that the practical information given here is helpful.

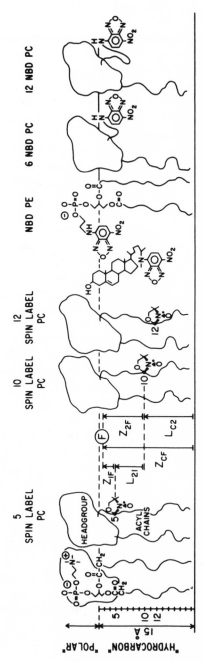

FIGURE 16. Schematic diagram of half of a membrane bilayer showing the structures of the spin-labeled PCs (5 SL PC, 10 SL PC, and 12 SL PC) and the NBD-labeled lipids (NBD PE, 6 NBD PC, 12 NBD PC, and NBD cholesterol). The vertical axis to the left represents the distance from the polar head groups of the lipids to the center of the bilayer. See the original article for definitions of Z and L parameters. Reprinted from Chattopadhay and London (1987) with permission from *Biochemistry*.

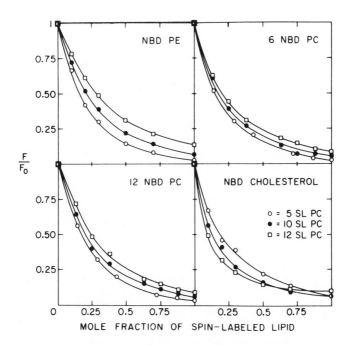

FIGURE 17. Fluorescence quenching of NBD-labeled lipids in phospholipid vesicles containing one of the three spin-labeled phospholipid: O, 5 SL PC; ●, 10 SL PC; □, 12 SL PC. Abscissa is the mole fraction of spin-labeled phospholipid in the lipid mixture. Reprinted from Chattopadhyay and London (1987) with permission from *Biochemistry*.

ACKNOWLEDGMENT. Some of the unpublished work presented here was supported by National Science Foundation Grant DMB 88-06113.

8. REFERENCES

Acuna, A. U., Lopez-Hernandez, F. J., and Oton, J. M., 1982, *Biophys. Chem.* **16**:253–260.

Altekar, W., 1977, *Biopolymers* **16**:341–368.

Alwattar, A. H., Lumb, M. D., and Birks, J. B., 1973, in *Organic Molecular Photophysics,* Vol. 1 (J. B. Birks, ed.), Wiley–Interscience, New York, pp. 403–454.

Ando, T., and Asai, H., 1980, *J. Biochem.* **88**:255–264.

Ando, T., Fujisaki, H., and Asai, H., 1980, *J. Biochem.* **88**:265–276.

Andre, J. C., Niclause, M., and Ware, W. R., 1978, *Chem. Phys.* **28**:371–377.

Badea, M. G., and Brand, L., 1979, *Methods Enzymol.* **61**:378–425.

Batenburg, A. M., Bougis, P. E., Rochat, H., Verkleij, A. J., and deKruijff, B., 1985, *Biochemistry* **24**:7101–7110.

Beechem, J. R., and Brand, L., 1985, *Annu. Rev. Biochem.* **54**:43–71.

Beechem, J. R., Ameloot, M., Knutson, J. R., and Brand, L., 1991, in *Fluorescence Spectroscopy: Theory and Applications,* Vol. 1 (J. R. Lakowicz, ed.), Plenum Press, New York.

Bieri, V. G., and Wallach, D. F. H., 1975, *Biochim. Biophys. Acta* **406**:415–423.

Blackwell, M. F., Gounairs, K., Zara, S. J., and Barber, J., 1987, *Biophys. J.* **51**:735–744.

Blatt, E., and Sawyer, W. H., 1985, *Biochim. Biophys. Acta* **822**:43–62.

Blatt, E., Chatelier, C., and Sawyer, W. H., 1986, *Biophys. J.* **50**:349–356.

Bohorquez, M. del V., Cosa, J. J., Garcia, N. A., and Previtali, C. M., 1984, *Photochem. Photobiol.* **40**:201–205.

Bushueva, T. L., Busel, E. P., and Burstein, E. A., 1975, *Stud. Biophys.* **52**:41–52.

Calhoun, D. B., Vanderkooi, J. M., and Englander, S. W., 1983, *Biochemistry* **22**:1533–1539.

Cavatorta, P., Favilla, R., and Mazzini, A., 1979, *Biochim. Biophys. Acta* **578**:541–546.

Chalpin, D. B., and Kleinfeld, A. M., 1983, *Biochim. Biophys. Acta* **731**:465–474.

Chattopadhayay, A., and London, E., 1987, *Biochemistry* **26**:39–45.

Chen, L. X. Q., Longworth, J. W., and Fleming, G. R., 1987, *Biophys. J.* **51**:685–873.

Chen, R. F., 1973, *Arch. Biochem. Biophys.* **158**:605–622.

Danileviciute, M., Adomeniene, O., and Dienys, G., 1981, *Org. React. (USSR)* **18**:217–224.

Demmer, D. R., James, D. R., Steer, D. P., and Verrall, R. E., 1987, *Biochemistry* **20**:4369–4377.

DeWolf, M. J. S., Van Dessel, G. A. F., Largrou, A. R., Hilderson, H. J. J., and Dierick, W. S. H., 1987, *Biochemistry* **26**:3799–3806.

Dowd, J. E., and Riggs, D. S., 1965, *J. Biol. Chem.* **240**:863–869.

Edwards, A. M. M., and Silva, E. S., 1986, *Radiat. Environ. Biophys.* **25**:113–122.

Eftink, M. R., 1991, in *Fluorescence Spectroscopy,* Vol. II, (J. R. Lakowicz, ed.), Plenum Press, New York (in press).

Eftink, M. R., and Ghiron, C. A., 1976a, *J. Phys. Chem.* **80**:486–493.

Eftink, M. R., and Ghiron, C. A., 1976b, *Biochemistry* **15**:672–680.

Eftink, M. R., and Ghiron, C. A., 1978, *Biochemistry* **7**:5546–5551.

Eftink, M. R., and Ghiron, C. A., 1981, *Anal. Biochem.* **114**:199–227.

Eftink, M. R., and Ghiron, C. A., 1984, *Biochemistry* **23**:3891–3899.

Eftink, M. R., and Ghiron, C. A., 1987a, *Photochem. Photobiol.* **45**:745–748.

Eftink, M. R., and Ghiron, C. A., 1987b, *Biochim. Biophys. Acta* **916**:343–349.

Eftink, M. R., and Hagaman, K. A., 1987, *Biophys. Chem.* **26**:277–282.

Eftink, M. R., and Selvidge, L. A., 1982, *Biochemistry* **21**:117–125.

Eftink, M. R., and Wasylewski, Z., 1989, *Biochemistry* **28**:382–391.

Eftink, M. R., and Wasylewski, Z., 1988b, *Biophys. Chem.* **32**:121–130.

Eftink, M. R., Zajicek, J. L., and Ghiron, C. A., 1977, *Biochim. Biophys. Acta* **491**:473–481.

Eftink, M. R., Selva, T. J., and Wasylewski, Z., 1987, *Photochem. Photobiol.* **46**:23–30.

Fato, R., Battino, M. D., Esposit, G. P., Castelli, G. P., and Lenaz, G., 1986, *Biochemistry* **25**:3378–3390.

Ferguson, B. Q., and Yang, D. C. H., 1986, *Biochemistry* **25**:529–539.

Froehlich, P. M., and Nelson, K., 1978, *J. Phys. Chem.* **82**:2401–2403.

Geisthardt, D., and Kruppa, J., 1987, *Anal. Biochem.* **160**:184–191.

Georghiou, S., Thompson, M., and Mukhopadhyay, A. K., 1982, *Biochim. Biophys. Acta* **688**:441–452.

Ghiron, C. A., Eftink, M. R., Porter, M. A., and Hartman, F. C., 1988, *Arch. Biochem. Biophys.* **260**:267–272.

Gratton, E., Alpert, B., Jameson, D. M., and Weber, G., 1984, *Biophys. J.* **45**:789–794.

Gomez-Fernandez, J. C., Baena, M. D., Teruel, J. A., Villalain, J., and Vidal, C. J., 1985, *J. Biol. Chem.* **260**:7168–7170.

Hansen, D., Altschmied, L., and Hillen, W., 1987, *J. Biol. Chem.* **262**:14030–14035.

Hashimoto, K., and Albridge, W. N., 1970, *Biochem. Pharmacol.* **19**:2591–2603.

Havel, H. A., Kauffman, E. W., and Elzinga, P. N., 1988, *Biochim. Biophys. Acta* **955**:154–163.

Hill, B. C., Horowitz, P. M., and Robinson, N. C., 1986, *Biochemistry* **25**:2287–2292.

Johnson, D. A., and Yguerabide, J., 1985, *Biophys. J.* **48**:949–955.

Kauffman, R. F., Chapman, C. J., and Pfeiffer, D. R., 1983, *Biochemistry* 22:3985–3992.

Kan, J. H., Wijnaendts van Resandt, R. W., and Dekkens, H. P. J. M., 1986, *J. Biomol. Struct. Dyn.* 3:827–842.

Knutson, J. R., Baker, S. H., Cappucino, A. G., Walbridge, D. W., and Brand, L., 1983a, *Photochem. Photobiol.* 37:S21.

Knutson, J. R., Beechem, J. M., and Brand, L., 1983b, *Chem. Phys. Lett.* 102:501–507.

Lakowicz, J. R., 1983, *Principles of Fluorescence Spectroscopy,* Plenum Press, New York, Chapter 9.

Lakowicz, J. R., Hogen, D., and Omann, G., 1977, *Biochim. Biophys. Acta* 471:401–411.

Lakowicz, J. R., and Weber, G., 1973a, *Biochemistry* 12:4161–4170.

Lakowicz, J. R., and Weber, G., 1973b, *Biochemistry* 12:4171–4179.

Lakowicz, J. R., Johnson, M. L., Joshi, N., Gryczynski, I., and Laczko, G., 1986, *Chem. Phys. Lett.* 131:343–348.

Lakowicz, J. R., Joshi, N. B., Johnson, M. L., Szmacinski, H., and Gryczynski, I., 1987, *J. Biol. Chem.* 262:10907–10910.

Lee, H.-J., Chen, Y.-H., and Chang, G.-G., 1988, *Biochim. Biophys. Acta* 955:119–127.

Lehrer, S. S., 1971, *Biochemistry* 10:3254–3263.

Lehrer, S. S., 1976, in *Biochemical Fluorescence: Concepts,* Vol. 2 (R. Chen and H. Edelhoch, eds.), Dekker, New York, pp. 515–544.

Lehrer, S. S., and Leavis, P. C., 1978, *Methods Enzymol.* 49:222–236.

London, E., and Feigenson, G. W., 1981, *Biochemistry* 20:1932–1938.

Markello, T., Zlotnick, A., Everett, J., Tennyson, J., and Holloway, P. W., 1985, *Biochemistry* 24:2895–2901.

Midoux, P., Wahl, P., Auchet, J.-C., and Monsigng, M., 1984, *Biochim. Biophys. Acta* 801:16–25.

Narasimhulu, S., 1988, *Biochemistry* 27:1147–1153.

Nemzek, T. L., and Ware, W. R., 1975, *J. Chem. Phys.* 62:477–489.

Olea, A. F., and Thomas, J. K., 1988, *J. Am. Chem. Soc.* 110:4494–4502.

O'Neil, K. T., Wolfe, H. R., Jr., Erickson-Viitanen, S., and DeGrado, W. F., 1987, *Science* 236:1454–1456.

Phillips, A. V., Robbins, D. J., Coleman, M. S., and Barkley, M. D., 1987, *Biochemistry* 26:2893–2903.

Pownall, H. J., and Smith, L. C., 1974, *Biochemistry* 13:2594–2597.

Ricci, R. W., and Kilichowski, K. B., 1974, *J. Phys. Chem.* 78:1953–1956.

Robbins, D. J., Deibel, M. R., Jr., and Markley, M. D., 1985, *Biochemistry* 24:7250–7257.

Ross, J. B. A., Schmidt, C., and Brand, L., 1981, *Biochemistry* 20:4369–4377.

Shinitzky, M., 1972, *J. Chem. Phys.* 56:5979–5981.

Shinitzky, M., and Rivnay, B., 1977, *Biochemistry* 16:982–986.

Sluyterman, L. A., and DeGraaf, M. J. M., 1970, *Biochim. Biophys. Acta* 200:595–597.

Sommers, P. B., and Kronman, M. J., 1980, *Biophys. Chem.* 11:217–232.

Steiner, R. F., and Kirby, E. P., 1969, *J. Phys. Chem.* 73:4130–4135.

Stryjewski, W., and Wasylewski, Z., 1986, *Eur. J. Biochem.* 158:547–553.

Stubbs, C. D., and Williams, B. W., 1991, in *Fluorescence Spectroscopy,* Vol. II (J. R. Lakowicz, ed.), Plenum Press, New York (in press).

Thulborn, K. R., and Sawyer, W. H., 1978, *Biochim. Biophys. Acta* 511:125–140.

Torgerson, P. M., 1984, *Biochemistry* 23:3002–3007.

Toulmé, J.-J., LeDuan, T., and Helene, C., 1984, *Biochemistry* 23:1195–1201.

Tyson, P. A., and Steinberg, M., 1987, *J. Biol. Chem.* 262:14030–14035.

Vaughn, W. M., and Weber, G., 1970, *Biochemistry* 9:464–473.

Wasylewski, Z., Koloczek, H., and Wasniowska, A., 1988, *Eur. J. Biochem.* 172:719–724.

Zinger, D., and Geacintov, N. E., 1988, *Photochem. Photobiol.* 47:181–188.

Chapter 2

Luminescent Trivalent Lanthanides in Studies of Cation Binding Sites

Matthew Petersheim

1. INTRODUCTION

Although the trivalent lanthanides are found at low levels throughout the biosphere and have been found to have some biological activity such as analgesia (Zuxuan *et al.*, 1985) and amino acid cotransport (Birnir *et al.*, 1987; Stevens and Kneer, 1988), there are no known metabolic processes in which they are essential. They are of interest in biochemistry for two general reasons. The lanthanides are a series of hard Lewis acids varying slightly in effective radius and, as such, are useful as steric probes of systems such as ion channels. Second, their paramagnetic and luminescent properties make them versatile spectroscopic probes of cation binding sites.

There are several comprehensive reviews of the theory of lanthanide paramagnetic properties (La Mar *et al.*, 1973; Reuben and Elgavish, 1979), optical properties (Wybourne, 1965; Dieke, 1968; Peacock, 1975; Reisfeld, 1975, 1976; Hüfner, 1978; Carnall, 1979; Blasse, 1979), and as probes of biochemical systems (Nieboer, 1975; Reuben, 1979; Horrocks and Sudnick, 1979; Richardson, 1982; Horrocks and Albin, 1984). The following is intended more as a survey than a thorough review and often the foregoing reviews will be referenced rather than primary sources in order to provide a wider perspective.

MATTHEW PETERSHEIM • Department of Chemistry, Seton Hall University, South Orange, New Jersey 07079.

As probes of cation binding sites, the luminescent lanthanides provide a variety of tools. Absorption and emission is often by way of 4f-to-4f transitions, but can also involve the empty 5d levels, charge transfer and energy transfer from nearby chromophores. Band position, shape, and intensity are all sensitive to the symmetry and composition of the inner coordination sphere and can be used to monitor changes in the lanthanide site. Binding site heterogeneity can be detected by the multiplicity of some bands and luminescence lifetime measurements. The symmetry of the ligand field can sometimes be partly characterized from progressions in band splitting and the composition of the inner sphere can be partly defined from positions of vibronic sidebands and the effect on luminescence lifetimes of isotopic substitutions in the ligands. Energy transfer experiments can be used to estimate the proximity of other chromophores to the lanthanide and analysis of the circular polarization of the absorption or emission bands can be used to establish the presence of a chiral environment.

Eu^{3+} has been used in a large fraction of the luminescence studies, often because it has a unique 4f-to-4f absorption/emission band near 580 nm which cannot be split by the ligand field (vide infra). Any evidence of other than a single band within 2 nm of 580 nm is a clear indication of binding site heterogeneity. Albin and Horrocks (1985) have also correlated the position of this band with the number of charged groups in the ligand field. The splitting of some of the other Eu^{3+} bands can be used to reject general categories of ligand field symmetries, and the luminescence decay time can be used to count the number of water molecules in the coordination sphere.

Usually a tunable dye laser is required to study the Eu^{3+} 580-nm band because of the resolution and sensitivity needed for analysis of the band shape (Horrocks and Sudnick, 1981; Horrocks and Albin, 1984; Albin and Horrocks, 1985; Herrmann *et al.*, 1986; Joshi and Shamoo, 1987). Eu^{3+} is also the least stable lanthanide with respect to redox chemistry. It has a reduction potential of about -0.35 V, depending on the coordination environment, which is within the range of weak biochemical reducing agents. All of the other trivalent lanthanides have oxidation/reduction potentials unfavorable by 1 V or more. Thus, for either instrumental or chemical reasons, it is not always possible to take advantage of some of the superior probe properties of Eu^{3+}.

Tb^{3+} has 4f-to-5d absorption bands (Carnall, 1979) which may provide an alternative to the demanding instrumental requirements of studying the 580-nm band of Eu^{3+}. Because the 5d orbitals are immediately exposed to the ligand field, the 4f-to-5d absorption bands move by thousands of wavenumbers with changes in coordination, providing a convenient monitor of site multiplicity and chemical exchange (Conti *et al.*, 1987; Halladay and Petersheim, 1988; Halladay, 1989; Petersheim and Sun, 1989; Petersheim *et al.,* 1989a,b). The baricenter and splitting of the $5d$ levels are sensitive to the nephelauxitic effect and ligand field symmetry, potentially providing information equivalent to that available from the position of the 580-nm band of Eu^{3+}. However, some of the Tb^{3+} absorption bands are too far into the ultraviolet to be studied with conventional fluorescence equipment and often overlap with ligand absorption bands. Energy transfer from intrinsic fluorophores, such as phenylalanine, tyrosine, tryptophan, and nucleic acids, generally gives rise to enhanced Tb^{3+} emission. This is often not the case with Eu^{3+} if charge transfer occurs (Martin and Richardson, 1979; Horrocks and Albin, 1984).

Excitation of Ce^{3+} occurs by way of allowed $4f$-to-$5d$ transitions that are several thousand wavenumbers (30 nm or so) farther toward the red than the $4f$-to-$5d$ bands of Tb^{3+}, making more of the excitation spectrum accessible on conventional fluorescence spectrometers. As mentioned above, the multiplicity and positions of these bands can be related to the strength and symmetry of the ligand field. Emission is also from the $5d$ excited state, providing a second source of ligand field parameters. The relative merits of this probe are being tested with phospholipid complexes (Sun and Petersheim, 1990; Sun, 1990).

The other lanthanides have spectral properties which may be suited to special problems. For example, in energy transfer experiments spectral overlap and even mechanism of transfer can be varied across the series. In order to limit the scope of this survey, the emphasis will be restricted to Tb^{3+} and Eu^{3+} studies with some indulgence in a personal interest in Ce^{3+}.

2. CHEMICAL PROPERTIES

2.1. Coordination Chemistry

Thompson (1979) and Sinha (1976a,b) provide comprehensive reviews of lanthanide coordination chemistry. Although crystal structures are available for complexes involving coordination numbers from 3 to 12, values from 6 to 10 seem to predominate (Horrocks and Albin, 1984). The coordination number and geometry of the complex are usually defined by the effective radius of the lanthanide and packing constraints on the ligands. There is virtually no directional preference imposed by the lanthanide.

Lanthanide complexes in aqueous solution are often very labile (Horrocks and Albin, 1984) and there are usually several possible geometries associated with a given coordination number which may differ only slightly in stability (Muetterties and Wright, 1967; Guggenberger and Muetterties, 1976; Sinha, 1976a). Thus, exchange among several forms of a complex may be relatively facile, unless constrained by the environment, and a search for some absolute solution structure may be futile. In crystal studies, this structural variability has been dubbed "promiscuous coordination" (Sinha, 1976a). The trivalent lanthanides preferentially bind to oxygen-containing ligands with coordination by nitrogen groups playing a secondary role especially in aqueous solutions (Sinha, 1976b; Thompson, 1979). Lanthanide(III)–ligand bonds are usually greater than 0.22 nm in length and described as predominantly ionic in nature. Slight covalent character in the bonds has been invoked to explain hypersensitivity in some electronic transitions (Henrie et al., 1976), anomalous behavior of some ionization potentials (Jørgensen, 1975), and the Fermi contact interaction observed in NMR experiments (Fischer, 1973; Jesson, 1973).

2.2. Chemical Trends across the Series

The ionic radii of the lanthanides decrease almost linearly across the series, spanning roughly a 0.02-nm range. There is also a linear increase in ionic radius with coordination number, e.g., La^{3+} changes from 0.1061 nm for a coordination number of 6 to 0.132 nm for 12 ligands (Sinha, 1976a,b). Since bonding is predominantly ionic, formation constants for lanthanide complexes might be expected to follow the smooth progression in ionic radius across the series. This is not observed often. Instead, the trend is disrupted in midseries in the vicinity of Eu^{3+}, Gd^{3+}, and Tb^{3+}, as depicted in Fig. 1A for complexation with trypsin (Epstein et al., 1974). Similar behavior has been observed for parvalbumin (Breen et al., 1985a,b). The history, evidence, and rationalization of the broken trend are discussed in detail by Sinha (1976b). In short, the stability of ionic lanthanide complexes depends on both effective radius and efficiency of screening the nuclear charge. Nuclear screening is a function of the total orbital angular momentum for the ground state, and the lanthanides fall into four categories:

- $L = 0$: La^{3+}, Gd^{3+}, and Lu^{3+}
- $L = 3$: Ce^{3+}, Eu^{3+}, Tb^{3+}, and Yb^{3+}
- $L = 5$: Pr^{3+}, Sm^{3+}, Dy^{3+}, and Tm^{3+}
- $L = 6$: Nd^{3+}, Pm^{3+}, Ho^{3+}, and Er^{3+}

This subdivision within the series can give rise to four linear correlations between the free energy change for complex formation and the L quantum number. These four correlations often follow the form of an "inclined W," as shown with trypsin (Fig. 1B).

The aquo-lanthanide(III) complexes have eight to ten coordinating water molecules, varying across the series (Horrocks and Sudnick, 1979;

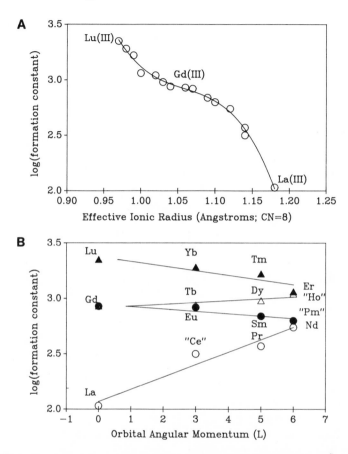

FIGURE 1. Variation in lanthanide affinity for porcine trypsin (Epstein *et al.,* 1974). (A) Nonlinear correlation between effective ionic radius and logarithm of the formation constant for the complex. The effective atomic radii are from Sinha (1976b) assuming a coordination number of 8, and the formation constants for Ce^{3+}, Pm^{3+} and Ho^{3+} were interpolated from the data of Epstein *et al.* (1974). (B) "Inclined W" from L-dependence of complex stability.

Horrocks and Albin, 1984). In a novel set of experiments, Kupke and Fox (1989) measured molar volume changes for a series of lanthanides binding to a calmodulin fragment and analogues of the fragment. An increase in solution volume is expected upon complexation due to release of directly bound water and electrostricted water molecules by both the trivalent cation and the polyanionic peptide. For structureless ions, the degree of electrostriction varies inversely with ionic radius. Assuming only the ionic radius changes across the lanthanides series, the volume change should increase smoothly in

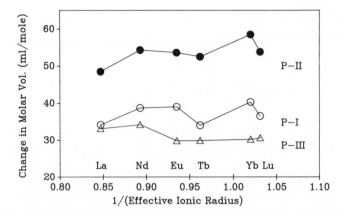

FIGURE 2. Molar volume changes resulting from complexation of calmodulin fragments with lanthanides (Kupke and Fox, 1989). The calmodulin fragments are: P-I, Ac-Asp-Ala-Asp-Gly-Asn-Gly-Thr-Ile-Asp-Phe-Pro-Glu-Phe-NH$_2$; P-II, Ac-Asp-Ala-Asp-Gly-Asp-Gly-Thr-Ile-Asp-Phe-Pro-Glu-Phe-NH$_2$; P-III, Ac-Asp-Ala-Asn-Gly-Asp-Gly-Thr-Ile-Asp-Phe-Pro-Glu-Phe-NH$_2$. The effective ionic radii are given in angstroms and were selected from the table provided by Sinha (1976b) assuming a coordination number of eight.

progressing from La^{3+} to Lu^{3+}. Figure 2 shows some of their results. The absence of a simple linear correlation in these plots is partly due to the L-dependence in the effective charge density on the cation. The pattern observed for the P-III fragment differs slightly from those for P-II and P-I. It may be that for P-III the lanthanide contribution to the volume change may be small compared with that from the peptide itself.

The conclusion that can be drawn from the observations in this section is that the lanthanides are very similar to each other in chemical behavior but are not identical. Ionic radius, effective charge, and preferred coordination number all change subtly across the series, and all three must be considered in a study of the relative properties of these cations.

3. SPECTROSCOPIC PROPERTIES OF THE LANTHANIDES

All of the trivalent lanthanides, except lanthanum itself, absorb light in the ultraviolet/visible spectrum and can emit light in the visible or near infrared with varying efficiencies. Details of the spectral properties for the entire series are given by Dieke (1968), Peacock (1975), Henrie *et al.* (1976), Hüfner (1978), Blasse (1979), and Carnall (1979), the last of whom provides absorption spectra for each lanthanide. Most luminescence studies with biochemical systems have involved either Eu^{3+} or Tb^{3+}, with occasional interest in the other lanthanides. Some unique properties of Ce^{3+} will be compared with those of the more common pair of probes.

3.1. Excitation Band Positions

The absorption spectra of aquo-complexes of Ce^{3+}, Tb^{3+}, and Eu^{3+} are presented in Fig. 3 with band assignments given in Table I. The aquo-Ce^{3+} spectrum consists solely of $4f \rightarrow 5d$ bands at 210, 222, 239, and 252 nm with a much weaker broad band at about 300 nm. Another strong band is reported to be just below 200 nm (Carnall, 1979). Although this band multiplicity has been interpreted in terms of ligand field splitting of the d levels, Svetashaev and Tsvirko (1984) found the temperature dependence of the bands to be indicative of more than one form of the aquo-Ce^{3+} complex. They also found that emission occurs from only one of the complexes, with excitation through the other forms occurring by way of rapid chemical interconversion. According to their study, at room temperature the emissive form represents about 2% of the total Ce^{3+}.

Carnall *et al.* (1968a) identified 16 primary $4f \rightarrow 4f$ aquo-Tb^{3+} absorption bands between 200 and 500 nm (Table I), with another 21 bands which could also appear in the same region. All have extinction coefficients of less than 0.4 M^{-1} cm^{-1} and each of these bands can exhibit splitting of a couple hundred cm^{-1}, depending on the strength and symmetry of the ligand field. The spectrum in Fig. 3B was obtained with a photodiode array with only 2-nm resolution; consequently, some of the 15 bands reported in this region by Carnall *et al.* (1968a) are not resolved. In addition, the bands at 256 and 265 are obscured by a broad band at 263 nm (extinction coefficient = 0.7 M^{-1} cm^{-1}) arising from what is assumed to be a spin-forbidden $4f \rightarrow 5d$ transition (Carnall, 1979). The fully allowed $4f \rightarrow 5d$ band at 220 nm (extinction coefficient = 310 M^{-1} cm^{-1}) completely obscures any other $4f \rightarrow 4f$ bands in this region (Fig. 3B).

The $5d$ levels of Ce^{3+} and Tb^{3+} are directly exposed to the ligand field and can be expected to change by thousands of wavenumbers with variation in the ligand field while the $4f$ levels are shielded by the full $5s^2$ and $5p^6$ orbitals. As a consequence, the $4f$ energy levels are perturbed by the ligand field on the order of only a few hundred cm^{-1} or less.

As mentioned before, Svetashaev and Tsvirko (1984) attributed the multiplicity of $4f \rightarrow 5d$ absorption bands for aquo-Ce^{3+} to coexisting forms of the complex. There is no equivalent evidence for multiple forms of the aquo-Tb^{3+} complex. Only a single allowed $4f \rightarrow 5d$ band is observed with aquo-Tb^{3+}, although there may be other bands at energies inaccessible on conventional absorption instruments. Also, luminescence decay for aquo-Tb^{3+} is fit

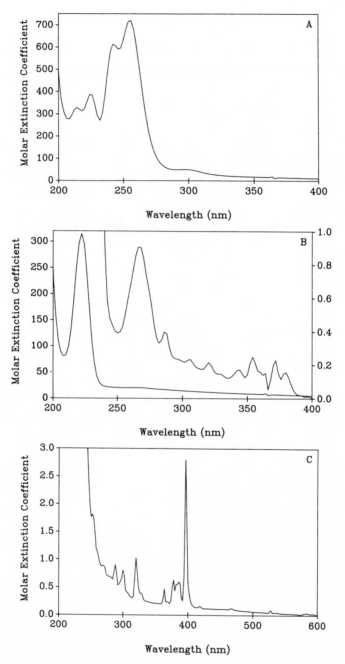

FIGURE 3. Absorption spectra of aquo-complexes. (A) Cerium (III) chloride in water. (B) Terbium (III) chloride in water; the left-hand ordinate axis is for the spectrum with a single dominant band. (C) Europium (III) perchlorate prepared by dissolving Eu_2O_3 in a stoichiometric amount of hot perchloric acid. The strong band offscale to the left is due to contaminating Eu^{2+}. Note that the $^5D_0 \leftarrow {}^7F_0$ band at 580 nm has an amplitude about the thickness of the line on this scale. All spectra were collected on a photodiode array spectrometer with 2-nm resolution.

TABLE I. Energy Level Assignments for Aquo-Ce^{3+}, -Tb^{3+}, and -Eu^{3+}

Aquo-Eu(III)[a]			Aquo-Tb(III)[b]			Aquo-Ce(III)[c]		
	cm^{-1}	nm^d		cm^{-1}	nm^d		cm^{-1}	nm
$^7F_0$	0		$^7F_6$	85		$^2F_{5/2}$	0	
$^7F_1$	360		$^7F_5$	2100		$^2F_{7/2}$	2235	
$^7F_2$	1020		$^7F_4$	3356		$5d$	33,333	300
$^7F_3$	1887		$^7F_3$	4400		$5d$	39,620	252
$^7F_4$	2865		$^7F_2$	5028		$5d$	41,841	239
$^7F_5$	3908		$^7F_1$	5440		$5d$	45,004	222
$^7F_6$	4980		$^7F_0$	5700		$5d$	47,619	210
$^5D_0$	17,277	579	$^5D_4$	20,500	490			
$^5D_1$	19,028	526	$^5G_6$	26,500	379			
$^5D_2$	21,519	465	$^5L_{10}$	27,100	370			
$^5D_3$	24,408	410	$^5G_5$	27,800	361			
$^5L_6$	25,400	394	$^5L_9$	28,400	353			
$^5G_2$	26,300	380	$^5L_8$	29,300	342			
$^5G_4$	26,620	376	$^5L_7$	29,450	341			
$^5G_6$	26,700	375	$^5D_1$	30,650	327			
$^5D_4$	27,670	361	$^5H_7$	31,600	317			
$^5H_4$	31,250	320	$^5H_6$	33,000	304			
$^5H_6$	31,520	317	$^5H_5$	33,900	295			
$^5F_2$	33,190	301	$^5F_5$	34,900	287			
$^5F_4$	33,590	298	$^5I_8$	35,200	285			
?	35,030	285	$^5I_7$	36,700	273			
$^5K_5$	36,205	276	$^5I_6$	37,760	266			
$^5K_6$	37,440	267	$^5K_9$	39,100	256			
?	39,060	256	$5d^*$	38,000	263			
$^5G_4$	39,890	251	$5d$	45,540	220			
$^5G_6$	41,370	242						

[a] Observed absorption transitions from Carnall *et al.* (1968b). A more complete set of assignments including the missing term symbols is provided in that reference.
[b] Observed absorption transitions from Carnall *et al.* (1968a). A more complete set of assignments including the missing term symbols is provided in that reference.
[c] Observed absorption transitions from Svetashaev and Tsvirko (1984) and Carnall (1979).
[d] Absorption from the lowest ground state.
* Spin-forbidden $4f$-to-$5d$ transition.

well with a single exponential function, although this is inconclusive since chemical exchange of the water molecules is faster by a factor of about 10^5 (Horrocks and Sudnick, 1979). Likewise, aquo-Eu^{3+} presents no clear evidence of multiple forms. Horrocks and Sudnick (1979) have presented evidence indicating that aquo-Ce^{3+} falls in the category of lanthanides that average ten coordinating water molecules while Tb^{3+} and Eu^{3+} appear to have nine. The higher coordination number of Ce^{3+} may be responsible for the multiplicity of structures reported by Svetashaev and Tsvirko (1984), which are apparently absent with Tb^{3+} and Eu^{3+}. It is equally likely that there are several forms of the aquo-Tb^{3+} and aquo-Eu^{3+} complexes that are simply more elusive than those of aquo-Ce^{3+}.

The $^2F_{7/2}$ Ce^{3+} state and the $^7F_5$ Tb^{3+} state are both about 2000 cm^{-1} above their respective ground states, which is too high for these levels to be appreciably populated at room temperature ($kT/hc = 207$ cm^{-1}). Thus, for these two lanthanides absorption is from the lowest ground state established by Hund's rules. The $^7F_1$ state of Eu^{3+} is only about 360 cm^{-1} above the $^7F_0$ ground state and excitation from both levels is observed. Unlike Ce^{3+} and Tb^{3+}, the Eu^{3+} $5d$ levels lie well above 50,000 cm^{-1} and no $4f \rightarrow 5d$ absorption bands are observed above 200 nm. Like Tb^{3+}, Eu^{3+} has many $4f$ levels accessible within the visible and ultraviolet regions yielding a large number of absorption and emission bands (Table I). The strongest of these can be seen in the aquo-Eu^{3+} spectrum of Fig. 3C. The very strong absorbance growing in below 300 nm is due to a $4f \rightarrow 5d$ band (extinction coefficient $= 2000$ M^{-1} cm^{-1}) from some Eu^{2+} inadvertently produced in preparing this sample from Eu_2O_3 and perchloric acid. Eu^{3+} is the least stable trivalent lanthanide, with a reduction potential of -0.35 V (Carnall, 1979). This is low enough to be affected by mild reducing agents such as $NADH_2$ at pH 7. The oxidation and reduction potentials for all of the other lanthanides are significantly greater than this. Because of the strong absorbance by Eu^{2+} and the sensitivity of the $4f \rightarrow 5d$ band position to changes in ligand field, a low level of Eu^{2+} contaminant could complicate the interpretation of excitation spectra in many biochemical complexes. Eu^{3+} also exhibits ligand-to-metal charge transfer bands which, in simple terms, is a consequence of the $4f^6$ configuration being one electron short of half-full. The $4f^8$ configuration of Tb^{3+} and $4f^1$ of Ce^{3+} make charge transfer rare (Blasse, 1979).

Eu^{3+} is unique among the lanthanides in that it has a nondegenerate ground state, $^7F_0$. Consequently, splitting of the absorption bands originating from this level reflects the effect of the ligand field on only the excited state. Excitation into the $^5D_0$ level, the lowest-lying excited state, provides a single band for each chemically distinct form of the complex. This transition has been thoroughly exploited as a means for determining the number of Eu^{3+} coordination states in a variety of chemical and biochemical systems (Horrocks and Sudnick, 1979; Rhee *et al.*, 1981; Snyder *et al.*, 1981; Horrocks and Albin, 1984; Breen *et al.*, 1985a,b; Albin and Horrocks, 1985; Herrmann *et al.*, 1986; Joshi and Shamoo, 1987; Hofmann *et al.*, 1988). As alluded to above, the aquo-Eu^{3+} spectrum appears to have a single $^5D_0 \leftarrow {}^7F_0$ band at 578.9 nm, suggesting that it does not have the multiple forms proposed for aquo-Ce^{3+}. This does not rule out the possibility that bands for multiple aquo-Eu^{3+} forms are simply not resolved.

In principle, the ligand field can be categorized as having cubic, hexagonal, tetragonal, or lower symmetry based on the multiplicity of bands for the $^5D_1 \leftrightarrow {}^7F_0$, $^5D_2 \leftrightarrow {}^7F_0$, $^5D_3 \leftrightarrow {}^7F_0$, and $^5D_4 \leftrightarrow {}^7F_0$ transitions (Horrocks and Albin, 1984). A more detailed description of the selection rules and their limitations is provided by Hüfner (1978). Albin and Horrocks (1985) have also shown that the position of the $^5D_0 \leftarrow {}^7F_0$ band follows a quadratic dependence on total ligand charge in the vicinity of the cation.

3.2. Emission Band Positions

Ce^{3+} is unique among the lanthanides in that emission involves the $4f^0 5d^1 \rightarrow 4f^1$ transition. This is fully allowed and, as a result, occurs on a nanosecond time scale (Blasse, 1979). The emission spectrum of aquo-Ce^{3+} consists of only two bands, one at 370 nm skewed toward lower wavelengths and another with a maximum at about 685 nm and a shoulder at about 765 (Fig. 4A). The shoulders are evidence of transitions to both the $^2F_{5/2}$ and $^2F_{7/2}$ ground states (Blasse, 1979) and the band separation is about 14,000 cm^{-1}, which is reasonable for ligand field splitting of the $5d$ $^2D_{3/2}$ excited state (Ryan and Jørgensen, 1966; Blasse, 1979). These same two emission bands are observed regardless of which of the excitation bands is chosen, which is consistent with the findings of Svetashaev and Tsvirko (1984) that only one form of the complex is emissive.

Although Tb^{3+} excitation can involve the $5d$ levels, there is rapid internal conversion of all excited states to the $4f^5 D_4$ level, from which virtually all emission occurs (Horrocks and Albin, 1984). There are seven ^7F_J states into which the $^5D_4$ can decay, giving relatively narrow bands at roughly 490, 545, 590, 620, 650, 670, and 685 nm for $J = 6$ to 0, respectively (Fig. 4B). Each of these emission bands can be split or shifted by the ligand field over a range of 10 nm as shown in Fig. 5 with the $^5D_4 \rightarrow {}^7F_5$ band for aquo-Tb^{3+} and complexes with two different phospholipids.

Emission from the $^5D_0$ state of Eu^{3+} yields seven bands: 579, 590, 620, 650, 690, 741, and 805 nm for transitions to $^7F_0$ through $^7F_6$, respectively, as was observed with Tb^{3+}. The Eu^{3+} emission spectrum also contains bands from the $^5D_1$ excited state which are lower in intensity and shifted to higher energy by about 1700 cm^{-1}. Equivalent emission from the $^5D_3$ excited state of Tb^{3+} is not as readily observed. The strongest band in the aquo-Eu^{3+} emission spectrum is from the $^5D_0 \rightarrow {}^7F_1$ transition near 591 nm (Fig. 4C). This is a magnetically allowed transition and its dominance in the spectrum is indicative of a highly symmetric ligand field (Hüfner, 1978; Blasse, 1979). In complexes of lower symmetry, the $^5D_0 \rightarrow {}^7F_2$ band at about 615 nm is usually more intense. The amplitude of the 615-nm band is "hypersensitive"

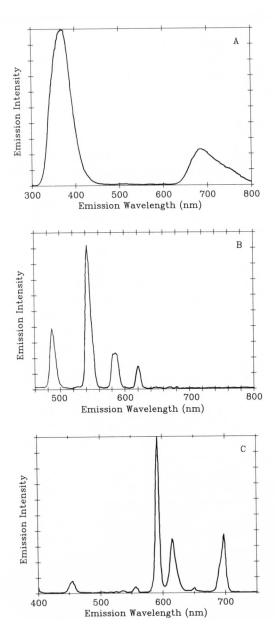

FIGURE 4. Emission spectra for aquo-complexes. (A) Cerium (III) chloride, 5×10^{-6} M, with excitation at 257 nm. The very low-amplitude band at 514 nm is due to second-order diffraction by the emission grating. (B) Terbium (III) chloride, 2×10^{-3} M, with excitation at 263 nm. (C) Europium (III) perchlorate, 8×10^{-3} M, with excitation at 394 nm.

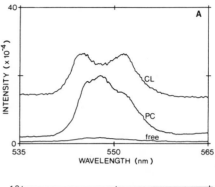

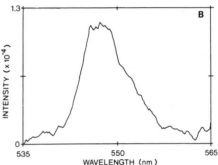

FIGURE 5. Structure of the Tb^{3+} 545-nm emission band. (A) Aquo-Tb^{3+} (2.4×10^{-4} M). (B) The spectrum labeled "CL" is 5×10^{-4} M bovine cardiolipin (diphosphatidylglycerol) with 2.4×10^{-4} M Tb^{3+}; that labeled "PC" is 3.8×10^{-3} M egg phosphatidylcholine with 2.4×10^{-4} M Tb^{3+}. That labeled "free" is the same as the aquo-Tb^{3+} spectrum drawn to scale in order to show the more intense emission from the lipid complexes.

to the symmetry and composition of the ligand field and is useful in monitoring changes in complexation (Richardson, 1982; Horrocks and Albin, 1984). The $^5D_0 \rightarrow {}^7F_0$ transition near 579 nm is too low in amplitude to be clearly visible in this emission spectrum.

3.3. General Statements about Band Intensities

Excitation into the $5d$ levels from the $4f$ ground states is allowed because there is a change in wavefunction parity with the change in principal quantum number (Laportes's rule). Consequently, these absorption bands have extinction coefficients on the order of 10^2–10^3 cm^{-1} M^{-1}, as shown for Ce^{3+}, Tb^{3+}, and divalent Eu^{2+} (Fig. 3). In some complexes, ligand-to-Eu^{3+} charge transfer can occur, resulting in a broad, featureless absorption band with an amplitude comparable to the $4f$-to-$5d$ bands observed with Tb^{3+}, Pr^{3+}, and Ce^{3+} (Blasse, 1979).

The $4f$-to-$4f$ transitions are electric dipole forbidden because there is no change in wavefunction parity (Laportes's rule). Despite this, many lanthanide absorption bands have extinction coefficients of up to 1 or more cm^{-1}

M^{-1}, e.g., the band near 400 nm for aquo-Eu^{3+}. The most general explanation for this slight allowed character is that the ligand field induces mixing of the $4f$ states with higher levels having different principal quantum numbers, usually the $5d$ levels. This mixing can be accomplished with either a ligand field having no center of symmetry, i.e., static coupling, or by way of odd symmetry vibrations in the ligand field. For both, the mechanism is referred to as a forced electric dipole transition and is discussed in detail by Peacock (1975), Hüfner (1978), and Carnall (1979). Reid and Richardson (1985) describe a method for developing spectrum–structure relationships from the intensities of $4f \rightarrow 4f$ electric dipole transitions.

Some bands are referred to as hypersensitive because their intensity (oscillator strength) increases by as much as a factor of 200 in going from a centrosymmetric complex to one with no center of symmetry (Peacock, 1975; Henrie *et al.*, 1976; Hüfner, 1978). Of the three lanthanides discussed here, only Eu^{3+} has transitions considered to be hypersensitive: the $^5D_1 \leftarrow$ $^7F_1$ (535 nm) and $^5D_2 \leftarrow {}^7F_0$ (465 nm) absorption bands and the $^5D_0 \rightarrow {}^7F_2$ (620 nm) emission band. There are several mechanisms which may be operative in determining hypersensitivity and the theory is discussed in detail by Peacock (1975) and Henrie *et al.* (1976). Qualitatively, hypersensitive transitions provide a sensitive monitor of changes in coordination.

Transitions involving a change in J of ± 1 or 0 (except $0 \leftrightarrow 0$ transitions) are magnetic dipole allowed. L and S are supposed to remain unchanged in the transition, but the effective states of the lanthanides are linear combinations of Russell–Saunders states of the same J value. Consequently, selection rules pertaining to L and S are often relieved for the most part. These magnetic dipole allowed transitions are less intense than fully allowed electric dipole transitions by a factor of $10^4 - 10^6$ and are rarely greater than the stronger forced electric dipole transitions. They are usually significant only in high-symmetry complexes where the forced electric dipole mechanisms are not operative and can be used to confirm a high-symmetry species. The Eu^{3+} $^5D_1 \leftrightarrow {}^7F_0$ transitions are known to involve predominately a magnetic dipole mechanism, as mentioned before.

3.4. Lanthanide(III)–Lanthanide(III) Energy Transfer and Distance Measurements

The theory developed by Foerster (1948) and Dexter (1953) revealed the potential for nonradiative energy transfer experiments in determining distances between donor and acceptor fluorophores. Because most of the optical transitions of the lanthanides are Laportes forbidden, except for forced dipole contributions, the electric dipole character of the transitions is reduced to roughly the same magnitude as the electric dipole–quadrupole, electric quadrupole, magnetic dipole, and electron exchange contributions (Reisfeld, 1976; Hüfner, 1978; Blasse, 1979). As a result, it is often not possible to establish a dominant mechanism and the corresponding distance dependence for an energy transfer process. This is further complicated by the substantial mixing of Russell–Saunders states that occurs with the lanthanides, degrading the selection rules for the various mechanisms. Reisfeld (1976) and Hüfner (1978) provide a thorough overview of this problem.

If the donor and acceptor are within 0.4 nm, the electron exchange mechanism is likely to be important (Blasse, 1979). The magnetic dipole mechanism is probably not important if the transitions involved are not magnetically allowed ($S = 0$; $L = 0, \pm 1$; $J = 0, \pm 1$), although state mixing introduces some uncertainty in this statement. Often it is assumed that neither of these mechanisms is significant and lanthanide energy transfer is interpreted in terms of the electric dipole–dipole or electric dipole–quadrupole mechanism (Reisfeld, 1976; Hüfner, 1978). Hüfner points out that even if an electric multipole mechanism can be assumed, it is difficult to distinguish dipole–dipole, dipole–quadrupole, and quadrupole–quadrupole mechanisms using only time-resolved emission. For a single electric multipole mechanism, the emission intensity is given as:

$$I(t)/I(0) = \exp\{-tk_A - (C_B/C_0)\Gamma(1 - 3/s)(tk_A)^{3/s}\} \qquad (1)$$

where

$\quad k_A$ = decay constant for the donor
$\quad C_B$ = acceptor concentration
$\quad C_0$ = critical concentration
$\quad\quad = 3/4\pi R_0^3$
$\quad R_0$ = critical radius or distance for which the rate = k_A
$\quad\quad s$ = exponent for the distance dependence
$\quad\quad\quad$ = 6, 8, or 10 for dipole–dipole, dipole–quadrupole, and quadrupole–quadrupole, respectively
$\Gamma(\ \ )$ = gamma function

The shape of the decay curve is not very sensitive to the value of s, as shown in Fig. 6.

In theory it should be possible to distinguish the various mechanisms using a judicious choice of donor/acceptor spectral overlap. For example, the rate constant for the electric dipole–dipole mechanism relative to the electric dipole–quadrupole mechanism is (Reisfeld, 1976):

$$\frac{k_{dd}}{k_{dq}} = \frac{2R^2}{3} \frac{\{O_{2A}\langle J_A|U_2|J_A'\rangle^2 + O_{4A}\langle J_A|U_4|J_A'\rangle^2 + O_{6A}\langle J_A|U_6|J_A'\rangle^2\}}{\langle r_A^2\rangle^2\langle C_2\rangle^2\langle J_A|U_2|J_A'\rangle^2}$$

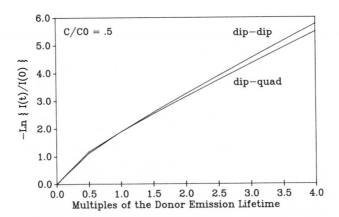

FIGURE 6. Comparison of hypothetical electric dipole–electric dipole and electric dipole–electric quadrupole emission decay profiles. The two curves were calculated from the natural logarithm of Eq. (1) in the text using discrete values of tk_A (multiples of the donor emission lifetime) and an acceptor concentration of half the critical concentration (C/C_0 = 0.5). For the electric dipole–electric dipole mechanism (dip–dip), "s" was set to 6 and for the electric dipole–electric quadrupole mechanism (dip–quad) "s" was set to 8. Note that in a real case, the two mechanisms would probably have different critical concentrations that are unknown parameters. The usual method for determining C_0 when only the dip–dip mechanism is (assumed) operative is invalid if one or more of the other mechanisms is significant.

where

$$R = \text{donor–acceptor separation}$$
$$O_{nA} = \text{Judd–Ofelt intensity parameter for acceptor}$$
$$\langle J_A | U_n | J_A' \rangle = \text{transition matrix element for acceptor excitation from state } J_A \text{ to state } J_A'$$
$$\langle r_A^2 \rangle = \text{radial integral for the } 4f \text{ electrons}$$
$$\langle C_2 \rangle = \text{"orientation factor"}$$

Each acceptor excitation band has a unique set of Judd–Ofelt parameters and transition matrix elements. The donor can be varied to select different excitation bands and, accordingly, vary the contributions from the two mechanisms. However, the values for the Judd–Ofelt parameters and transition matrix elements must be known in order to experimentally separate the contributions from the various mechanisms. Although these parameters can be estimated with reasonable accuracy, the ligand field must be well characterized, which is usually not the case in biochemical complexes.

In practice, it may be safe to assume that if the separation of the donor and acceptor is greater than 10 Å, the dominant mechanism is likely to be Foerster electric dipole–dipole with the familiar R^{-6} dependence (Reisfeld,

1976; Blasse, 1979; Snyder *et al.*, 1981). Even this statement does not appear to hold well for some pairs of lanthanides, e.g., Tb^{3+}/Nd^{3+} (Rhee *et al.*, 1981; Snyder *et al.*, 1981). At shorter distances there may be no single dominant mechanism that can be assumed or easily demonstrated with confidence.

3.5. Ligand-Sensitized Emission

One of the earliest studies employing lanthanides as luminescent biochemical probes demonstrated enhanced Tb^{3+} emission due to energy transfer from tyrosines of erythrocyte membrane proteins (Mikkelsen and Wallach, 1974). This form of enhanced lanthanide emission has been observed in several systems, many of which are discussed by Brittain *et al.* (1976a), Martin and Richardson (1979), Reuben (1979), De Jersey *et al.* (1981), Horrocks and Sudnick (1981), Richardson (1982), and Horrocks and Albin (1984). Tb^{3+} has been used as the acceptor in many of these experiments because it consistently demonstrates efficient emission enhancement. However, the allowed $4f \rightarrow 5d$ excitation band of Tb^{3+} has been observed to shift to as high as 300 nm in phospholipid complexes (Halladay and Petersheim, 1988; Petersheim *et al.*, 1989b). This should be considered when analyzing the excitation spectrum to establish the identity of the donor fluorophores. Ce^{3+} and Pr^{3+} are likely to present the same problem. Inadvertent reduction of Eu^{3+} to Eu^{2+} or the presence of ligand-to-metal charge transfer bands could also present a problem (Richardson, 1982).

In the aforementioned energy transfer studies, the ligand fluorophores were usually remote relative to the bound lanthanide and energy transfer was assumed to be electric dipole in nature. Several classes of ligands have chromophores more immediate to the lanthanide binding site. It has been found that many of these undergo rapid intersystem crossing to a triplet excited state followed by very efficient energy transfer to a coordinated lanthanide. This form of energy transfer has been called sensitized emission and can lead to an increase in luminescence by several orders of magnitude. Horrocks and Albin (1984) provide a thorough review of this topic.

3.6. Circularly Polarized Luminescence

When the environment of the lanthanide is chiral, the electronic transitions can exhibit optical activity. In order to avoid confusion with the circular dichroism of the ligands, the optical activity of the lanthanide center is usually measured from the polarization of the emission bands. This emission technique is often referred to as circularly polarized luminescence (CPL) and has been used to characterize biochemical complexes primarily by Richardson, Brittain, and co-workers (Brittain *et al.,* 1976b; Richardson, 1982; Horrocks and Albin, 1984). An example of a Tb^{3+}–phospholipid CPL spectrum is given in Fig. 7.

The magnitude of the CPL effect, as with circular dichroism, depends on the scalar product of the electric and magnetic transition dipoles. Strong CPL effects are observed with the $^5D_4 \rightarrow {}^7F_5$ (545 nm) band of Tb^{3+} and the $^5D_0 \rightarrow {}^7F_1$ (590 nm) band of Eu^{3+} (Richardson, 1980, 1982). Note that both have a change of spin of $\frac{1}{2}$, so neither is a formally allowed magnetic dipole transition.

CPL provides unambiguous evidence of a chiral environment for the lanthanide, although the chiral ligands need not be part of the inner coordination sphere. Optical activity induced by the outer coordination sphere is referred to as the Pfeiffer effect and has been demonstrated in association of chiral species with stable lanthanide chelates (Brittain, 1980, 1981; Madras and Brittain, 1980; cf. Horrocks and Albin, 1984). The sign on the Pfeiffer effect has been correlated with the absolute configuration of the chiral outer sphere ligand, providing a useful technique for enantiomer identification (Sen *et al.,* 1981; Yan *et al.,* 1982; Morley *et al.,* 1982).

4. BIOCHEMICAL APPLICATIONS

Cation binding to proteins, lipids, nucleic acids, and other biomolecules may be important either in terms of a nonspecific change in the electrostatic

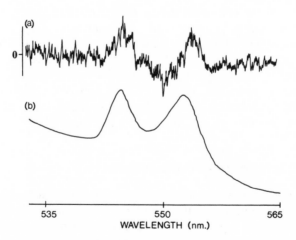

FIGURE 7. Circularly polarized luminescence spectrum (top) of the 545-nm emission band (bottom) for a Tb^{3+}–cardiolipin complex. The CPL spectrum was collected on an instrument constructed by H. G. Brittain.

properties of the system or specific stereochemical interactions resulting from complexation of the cation. Only the latter will be considered in the following discussion. Lanthanide luminescence can be used to determine the number of chemically unique binding sites, to characterize the coordination environment in each site, and to estimate distances between binding sites or from the cation to fluorescent groups within the biomolecule. A few representative examples of these studies are presented here.

4.1. Cation-Binding Site Heterogeneity

There are several calcium binding proteins with multiple binding sites that are well characterized and have been used to demonstrate the potential of the luminescent lanthanides as probes of the cation's environment. For example, the Eu^{3+} $^5D_0 \leftarrow {}^7F_0$ excitation band has been used to distinguish the multiple cation sites in parvalbumin (Rhee *et al.*, 1981), thermolysin (Snyder *et al.*, 1981), calbindin (Hofmann *et al.*, 1988), and calmodulin (Mulqueen *et al.*, 1985; Horrocks and Tingey, 1988). With some chemical systems there is insufficient resolution of the $^5D_0 \leftarrow {}^7F_0$ band to unambiguously determine the number of dominant binding sites. This limitation can be overcome, sometimes, by luminescence decay experiments. In a study of cation binding sites on intact sarcoplasmic reticulum membranes, Joshi and Shamoo (1987) observed multiexponential decay of Eu^{3+} emission. They interpreted their results in terms of two high-affinity sites that were not adequately resolved in the Eu^{3+} $^5D_0 \leftarrow {}^7F_0$ band. Likewise, Herrmann *et al.* (1986) detected multiple forms of Eu^{3+} complexes with phospholipids using luminescence decay experiments.

Horrocks and Tingey (1988) have improved the differentiation of Eu^{3+} chemical states using time-resolved excitation spectra for the $^5D_0 \leftarrow {}^7F_0$ transition. They obtained individual spectra for three distinct forms of the Eu^{3+}–calmodulin complex that were not well resolved in the more conventional excitation spectrum.

The $5d$ levels of Tb^{3+} and Ce^{3+} (and Pr^{3+}) change by thousands of wavenumbers with variation in the ligand field, resulting in excitation spectra that exhibit gross changes with the coordination environment. Also, the fully allowed $4f$-to-$5d$ bands have extinction coefficients at least 100 times greater than the $4f$-to-$4f$ bands, relieving the sensitivity constraints of the Eu^{3+} studies. These properties of the Tb^{3+} excitation spectrum have been used to detect heterogeneity in the cation binding sites of phospholipid membranes (Conti *et al.*, 1987; Halladay and Petersheim, 1988; Petersheim *et al.*, 1989a,b; Petersheim and Sun, 1989).

The disadvantages with Tb^{3+} derive from the fact that the allowed $4f$-to-$5d$ bands are usually just barely within the near-ultraviolet region. With aquo-Tb^{3+}, the lowest energy band that is fully allowed occurs at 220 nm. Presumably there are other $4f$-to-$5d$ bands due to ligand field splitting of the $5d$ levels but they may be located beyond 150 nm, depending on the strength of the ligand field. Thus, the information from ligand field splitting of the $5d$ levels that would rival the information available from the Eu^{3+} $^5D_0 \leftarrow {}^7F_J$ transitions, is effectively inaccessible. In addition, the position of this lowest energy allowed band is often outside of the usable lamp profile of most fluorescence spectrometers or the ligands have absorption bands that obscure the Tb^{3+} bands.

The strong Ce^{3+} $4f$-to-$5d$ absorption bands are significantly more accessible than those of Tb^{3+} (and Pr^{3+}), with the first intense band for aquo-Ce^{3+} at 252 nm. Ryan and Jørgensen (1966) prepared octahedral $CeCl_6{}^{3-}$ and $CeBr_6{}^{3-}$ complexes in acetonitrile which exhibited single intense absorption bands at 330 and 343 nm, respectively. A second band for each due to the e_g manifold of the $5d$ orbitals is presumed to lie below 250 nm. There is also a redshift in the excitation spectrum of Ce^{3+} complexes with phosphatidic acid. The most intense accessible band occurs above 295 nm (Fig. 8). This band moves by more than 15 nm for the Ce^{3+}–DMPA complex with the degree of protonation of the phosphate group, while the Ce^{3+}–DOPA com-

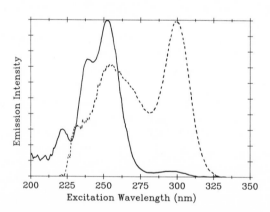

FIGURE 8. Excitation spectra for aquo-Ce^{3+} (solid line) and a Ce^{3+} complex with dioleoyl-phosphatidic acid (dashed line). The aquo-Ce^{3+} spectrum was obtained with 1×10^{-4} M $CeCl_3$ in water at 25°C and the lipid complex was prepared with 2.4×10^{-4} M dioleoyl-phosphatidic acid (DOPA), 1.5×10^{-5} M $CeCl_3$ in 0.1 M NaCl at 25°C and pH 6. Both spectra were collected with a low-ozone Xe arc lamp in the ratio mode to correct for the source profile and monitoring the emission at 375 nm. The aquo-Ce^{3+} excitation bands correspond in position and relative amplitudes to the absorption bands except that the Xe lamp intensity drops precipitously below 230 nm. There was insufficient lamp intensity to detect the 210-nm band. The Ce^{3+}–DOPA bands are at 256 and 301 nm. Their relative amplitudes do not change with temperature or the amount of Ce^{3+} bound, and both yield the same emission spectrum.

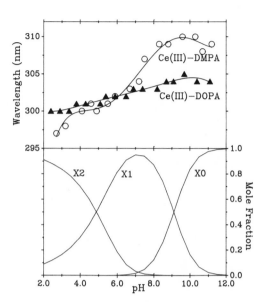

FIGURE 9. Redshift of the Ce^{3+}–dimyristoylphosphatidic acid (DMPA) and Ce^{3+}–DOPA excitation bands (top) with deprotonation of the phosphomonoester (bottom). The lipid was 5×10^{-4} M and Ce^{3+} was 2.5×10^{-5} M in 0.1 M NaCl at 25°C. The bottom graph is a plot of the mole fractions of the diprotonated (X2), monoprotonated (X1), and the aprotic (X0) forms of the lipid using the approach of Copeland and Andersen (1982) to account for the double layer effect. These results are presented in greater detail by Sun (1989) and Sun and Petersheim (1990).

plex undergoes a much smaller change in band position (Fig. 9). The two forms of phosphatidic acid differ in interfacial surface area, with DMPA packing being more sensitive to the degree of protonation (Patil *et al.*, 1979). Likewise, the Tb^{3+} excitation spectrum (Fig. 10) and Eu^{3+} emission spectrum (Fig. 11) show more extreme changes with the DMPA complexes than DOPA complexes. All three probes are reporting the same physical changes in the lipid, although the three lanthanides are not necessarily forming exactly the same complexes.

4.2. Counting Water Molecules and Characterizing the Coordination Sphere

Lanthanide luminescence is subject to quenching through vibronic coupling with inner sphere ligands. This presents the possibility of identifying ligands by observing the effect of isotopic substitution on the luminescence lifetime. In fact, there is a linear correlation between the lifetimes of Tb^{3+} and Eu^{3+} and the ratio of H_2O and D_2O in the inner sphere. The slope of the linear correlation is proportional to the number of bound water molecules (Horrocks and Albin, 1984). Horrocks and Sudnick (1979) demonstrated the reliability of this technique by comparing luminescence estimates with the number of coordinated water molecules observed in crystallographic studies of lanthanide–thermolysin complexes. This method for determining

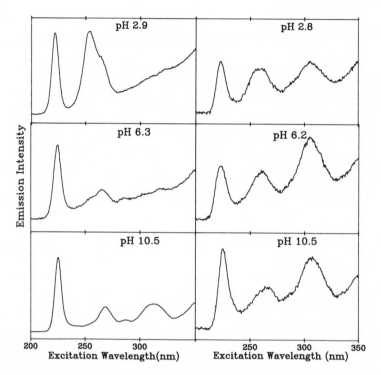

FIGURE 10. Changes in Tb^{3+}–phosphatidic acid coordination with the degree of proton-ation. At the left are excitation spectra for Tb^{3+}–DMPA complexes at pH 2.9 (top), 6.3 (middle) and 10.5 (bottom) and at the right are Tb^{3+}–DOPA excitation spectra at pH 2.8 (top), 6.2 (middle), and 10.5 (bottom). Note that with DMPA there are three distinctly different forms of the complex indicated by the disappearance of the band at 253 nm in the low-pH range and appearance of the 310-nm band in the high-pH range. The changes in the Tb^{3+}–DOPA spectra are much more subtle, as was observed with Ce^{3+}–DOPA. All spectra were collected with 2×10^{-3} M lipid, 1×10^{-4} M Tb^{3+} at room temperature, monitoring the emission at 545 nm. They were not corrected for the source intensity profile, which was a 1000-W Xe arc lamp through a prism monochromator. These results are presented in greater detail by Sun (1989) and Sun and Petersheim (1989).

the number of bound water molecules has become a standard procedure for those performing lifetime measurements.

In principle, the water count puts an upper limit on the likely number of coordination sites filled by other ligands. Horrocks and co-workers are at-tempting to extend the ligand count using the nephelauxitic shift of the Eu^{3+} $^{5}D_0 \leftarrow {}^{7}F_0$ band to estimate total ligand charge at the lanthanide (Albin and Horrocks, 1985). In a study of the four Ca^{2+} binding sites in calmodulin, Horrocks and Tingey (1988) found that Eu^{3+} in the high-affinity binding sites has two water molecules and an estimated total ligand charge of -2.0.

One of the weak binding sites has two water molecules and a charge of -2.5 and the other weak binding site has three waters and a charge of -2.9. These estimates of ligand charges are lower than expected from a sequence analysis of calmodulin, which places four carboxylates at one strong site and three carboxylates at each of the other sites. According to Horrocks and Tingey (1988), the charge discrepancy may be a consequence of some carboxylates being involved in bidentate coordination, which was not taken into account in the original study of the nephelauxitic trend (Albin and Horrocks, 1985). Another approach being explored involves the analysis of vibronic sidebands in the lanthanide excitation spectrum (Zolin *et al.*, 1975; Navon *et al.*, 1981; Iben *et al.*, 1989). Although the progress being made is exceptional, a complete accounting of the coordination environment is still not possible using luminescence studies alone.

In cases where weak acids or bases are involved as ligands, the coordination sphere can be indirectly probed by monitoring spectral changes as a function of pH (see Horrocks and Albin, 1984). Figures 9–11 show the spectral changes that occur with Ce^{3+}, Tb^{3+}, and Eu^{3+} bound to bilayer vesicles of phosphatidic acid as the phosphomonoester headgroup is deprotonated. Titrations of this type do not provide direct evidence for coordination by a particular functional group since a remote change in protonation may also induce a change in ligand packing.

Direct coordination by a functional group can be assumed if a Fermi contact interaction between the paramagnetic lanthanide and the ligand can be demonstrated in the ligand NMR spectrum (Jesson, 1973; La Mar, 1973; Fischer, 1973; Reuben, 1979). Identification of metal–ligand vibrational bands in infrared or Raman spectra would also be very useful in describing the coordination sphere. However, these bands occur at relatively low energy and are difficult to detect in aqueous solutions. As a result, there are little data of this type for biochemical complexes of lanthanides or other hard Lewis acids.

4.3. Energy Transfer and Distance Measurements

Horrocks and co-workers chose two proteins having multiple calcium binding sites for which the X-ray crystal structures were known with lanthanides substituted in the sites: parvalbumin (Rhee *et al.*, 1981) and thermolysin (Snyder *et al.*, 1981). A summary of their results is presented in Table II. The distances were calculated assuming an electric dipole–dipole transfer, which they demonstrate to be a valid assumption for all but the Tb^{3+}/Nd^{3+} pairs (Snyder *et al.*, 1981). The dominance of this mechanism may be a fortuitous consequence of the distance. Reisfeld (1976) suggests that for

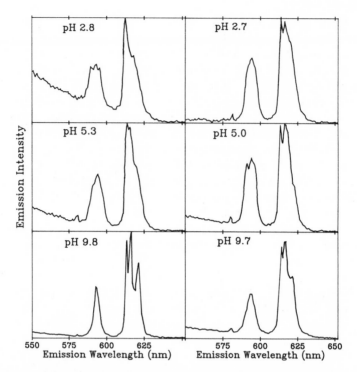

FIGURE 11. Changes in Eu^{3+}–phosphatidic acid coordination with the degree of protonation. At the left are emission spectra for Eu^{3+}–DMPA complexes at pH 2.8 (top), 5.3 (middle), and 9.8 (bottom) and at the right are Eu^{3+}–DOPA emission spectra at pH 2.7 (top), 5.0 (middle), and 9.7 (bottom). As with Ce^{3+} and Tb^{3+}, the changes in the Eu^{3+}–DMPA complex are clearer than those in the DOPA complex, although it is not as obvious that there are three different forms of the Eu^{3+}–DMPA species. Note that the $^5D_0 \rightarrow {}^7F_0$ band at about 580 nm is just barely detectable in all the spectra. All spectra were collected with 2×10^{-3} M lipid, 1×10^{-4} M Eu^{3+} in 0.1 M NaCl at room temperature with excitation at 395 nm. These results are presented in greater detail by Sun (1989) and Sun and Petersheim (1990).

short distances, i.e., less than 1 nm, the electric quadrupole–quadrupole mechanism may be more significant than the electric dipole–dipole mechanism. At greater distances the latter appears to be more important. Empirically, the Eu^{3+}/Nd^{3+} pair was most consistent with the X-ray data and was used in a later study of calmodulin sites also separated by more than 1 nm (Horrocks and Tingey, 1988).

Resonant energy transfer between aromatic amino acid residues and a lanthanide bound to a protein was first reported for transferrin by Luk (1971) and for cation binding sites on erythrocyte membranes by Mikkelsen and Wallach (1974). Examples of several other cation binding proteins ex-

TABLE II. Intramolecular Energy Transfer between Lanthanides

Parvalbumin[a]			Thermolysin[b]		
Donor	Acceptor	r (nm)[c]	Donor	Acceptor	r (nm)[c]
Eu^{3+}	Pr^{3+}	1.26	Eu^{3+}	Pr^{3+}	1.18
Eu^{3+}	Nd^{3+}	1.15	Eu^{3+}	Nd^{3+}	1.12
Tb^{3+}	Nd^{3+}	0.88	Tb^{3+}	ND^{3+}	0.86
Tb^{3+}	Ho^{3+}	1.01	Tb^{3+}	Ho^{3+}	1.15
Tb^{3+}	Er^{3+}	0.91	Tb^{3+}	Er^{3+}	1.10
Tb^{3+}	Pr^{3+}	0.94	Tb^{3+}	Pr^{3+}	1.09
X-ray		1.18[d]	X-ray		1.17[e]

[a] From Rhee *et al.* (1981).
[b] From Snyder *et al.* (1981).
[c] Separation between metal centers calculated from energy transfer efficiencies.
[d] Separation determined from X-ray diffraction by Sowadsky *et al.* (1978).
[e] Separation determined from X-ray diffraction by Matthews and Weaver (1974).

hibiting energy transfer to a bound lanthanide are given by Brittain *et al.* (1976a), Martin and Richardson (1979), Reuben (1979), De Jersey *et al.* (1981), and Horrocks and Albin (1984). Similar energy transfer studies with other aromatic biomolecules, such as porphyrins and nucleic acids, are reviewed by Reuben (1979) and Horrocks and Albin (1984).

Martin and Richardson (1979) suggest that the lanthanides act as isotropic acceptors in Foerster-type energy transfer from the aromatic groups, removing directional dependence from the R^{-6} relation. They also mention the possibility of an electron exchange mechanism dominating at distances approaching contact of the donor and the lanthanide. Horrocks and Collier (1981) demonstrated that the Foerster mechanism provides an accurate estimate of the distance from a tryptophan residue to Tb^{3+} substituted in parvalbumin. As in the interlanthanide energy transfer studies by the same group, the distance between donor and acceptor is greater than 1.0 nm (Horrocks *et al.*, 1980; Rhee *et al.*, 1981; Snyder *et al.*, 1981), which may be the reason why the other electric multipole and magnetic dipole mechanisms can be ignored. It remains to be demonstrated that the Foerster mechanism is dominant in energy transfer to lanthanides at shorter distances.

Intermolecular energy transfer between a lanthanide donor diffusing freely in solution and any type of acceptor is complicated by the millisecond luminescent lifetimes of the lanthanides (except Ce^{3+}). Any attempt to estimate distances between donor and acceptor must involve an average over the volume diffusively swept by the donor/acceptor pair, as discussed by Thomas *et al.* (1978). In subsequent studies of this "diffusion-enhanced" energy transfer, it was concluded that the dominant mechanism was the exchange interaction rather than electric dipole–dipole (Meares and Rice, 1981; Meares *et al.*, 1981; Meares and Wensel, 1984). Magnetic dipole and other electric multipole mechanisms were not considered.

5. PERSPECTIVES

So far, most of the work with the luminescent lanthanides has concentrated on characterization of binding sites. This will continue to be true with each new chemical system investigated. However, all of this effort is preparation for using the lanthanides to probe the action of cations such as Ca^{2+} in biochemical processes. Does cation binding simply provide a static stereochemical service or is the lability of hard Lewis acid interactions important in dynamic aspects of the structures formed? In systems having several different sites, which is important for a particular activity? The lanthanides provide a unique view of the cation's environment and sufficient spectroscopic versatility to address these questions even in complex mixtures.

The potential of the lanthanides as probes of biochemical events would be more fully realized if greater effort were made to adapt the probes to more conventional fluorescence equipment. For example, it would be useful to correlate the relatively detailed information gained from Eu^{3+} with spectral features of Tb^{3+} and Ce^{3+} in the same environment. Both are more easily monitored by a greater number of investigators since Tb^{3+} emission is more readily sensitized by intrinsic fluorophores and Ce^{3+} emission is more intense in the absence of sensitization. Demonstrating not only the power but also the availability of these tools will encourage a broader range of investigation into their use in the study of cation-induced events.

6. REFERENCES

Albin, M., and Horrocks, W. D., Jr., 1985, *Inorg. Chem.* **24**:895–900.
Birnir, B., Hirayama, B., and Wright, E. M., 1987, *J. Membr. Biol.* **100**:221–227.
Blasse, G., 1979, in *Handbook on the Physics and Chemistry of Rare Earths,* Vol. 4 (K. A. Gschneidner, Jr., and L. R. Eyring, eds.), North-Holland, Amsterdam, pp. 237–274.
Breen, P. J., Hild, E. K., and Horrocks, W. D., Jr., 1985a, *Biochemistry* **24**:4991–4997.
Breen, P. J., Johnson, K. A., and Horrocks, W. D., Jr., 1985b, *Biochemistry* **24**:4997–5004.
Brittain, H. G., 1980, *Inorg. Chem.* **19**:2136.
Brittain, H. G., 1981, *Inorg. Chem.* **20**:3007.
Brittain, H. G., Richardson, F. S., and Martin, R. B., 1976a, *J. Am. Chem. Soc.* **98**:8255–8260.
Brittain, H. G., Richardson, F. S., Martin, R. B., Burtnick, L. D., and Kay, C. M., 1976b, *Biochem. Biophys. Res. Commun.* **68**:1013–1019.

Carnall, W. T., 1979, in *Handbook on the Physics and Chemistry of Rare Earths,* Vol. 3 (K. A. Gschneidner, Jr., and L. R. Eyring, eds.), North-Holland, Amsterdam, pp. 171–208.

Carnall, W. T., Fields, P. R., and Rajnak, K., 1968a, *J. Chem. Phys.* **49**:4447–4449.

Carnall, W. T., Fields, P. R., and Rajnak, K., 1968b, *J. Chem. Phys.* **49**:4450–4455.

Conti, J., Halladay, H. N., and Petersheim, M., 1987, *Biochim. Biophys. Acta* **902**:53–64.

Copeland, B. R., and Andersen, H. C., 1982, *Biochemistry* **21**:2811–2820.

De Jersey, J., Jeffers-Morley, P., and Martin, R. B., 1981, *Biophys. Chem.* **13**:233–243.

Dexter, D. L., 1953, *J. Chem. Phys.* **21**:836.

Dieke, G. H., 1968, *Spectra and Energy Levels of Rare Earth Ions in Crystals,* Wiley, New York.

Epstein, M., Levitzki, A., and Reuben, J., 1974, *Biochemistry* **13**:1777.

Fischer, R. D., 1973, in *NMR of Paramagnetic Molecules* (G. N. La Mar, W. D. Horrocks, Jr., and R. H. Holm, eds.), Academic Press, New York, pp. 522–555.

Foerster, T., 1948, *Ann. Phys.* **2**:55.

Guggenberger, L. J., and Muetterties, I. L., 1976, *J. Am. Chem. Soc.* **98**:7221–7225.

Halladay, H. N., 1989, Ph.D. dissertation, Seton Hall University, South Orange, N.J.

Halladay, H. N., and Petersheim, M., 1988, *Biochemistry* **27**:2120–2126.

Henrie, D. E., Fellows, R. L., and Choppin, G. R., 1976, *Coordination Chemistry Reviews,* Vol. 18, Elsevier, Amsterdam, pp. 199–224.

Herrmann, T. R., Jayaweera, A. R., and Shamoo, A. E., 1986, *Biochemistry* **25**:5834–5838.

Hofmann, T., Eng, S., Lilja, H., Drakenberg, T., Vogel, H. J., and Forsen, S., 1988, *Eur. J. Biochem.* **172**:307–313.

Horrocks, W. D., Jr., and Albin, M., 1984, *Prog. Inorg. Chem.* **31**:1–104.

Horrocks, W. D., Jr., and Collier, W. E., 1981, *J. Am. Chem. Soc.* **103**:2856–2862.

Horrocks, W. D., Jr., and Sudnick, D. R., 1979, *J. Am. Chem. Soc.* **101**:335–340.

Horrocks, W. D., Jr., and Sudnick, D. R., 1981, *Acc. Chem. Res.* **14**:384–392.

Horrocks, W. D., Jr., and Tingey, J. M., 1988, *Biochemistry* **27**:413–419.

Horrocks, W. D., Jr., Rhee, M.-J., Snyder, A. P., and Sudnick, D. R., 1980, *J. Am. Chem. Soc.* **102**:3652–3653.

Hüfner, S., 1978, *Optical Spectra of Transparent Rare Earth Compounds,* Academic Press, New York.

Iben, I. E. T., MacGregor, R. B., Shyamsunder, E., Stavola, M., and Friedman, J. M., 1989, *Biophys. J.* **55**:519a (abstract W-Pos162).

Jesson, J. P., 1973, in *NMR of Paramagnetic Molecules* (G. N. La Mar, W. D. Horrocks, Jr., and R. H. Holm, eds.), Academic Press, New York, pp. 1–52.

Jørgensen, C. K., 1975, in *Structure and Bonding,* Vol. 22 (J. D. Dunitz, P. Hemmerich, R. H. Holm, J. A. Ibers, C. K. Jørgensen, J. B. Neilands, D. Reinen, and R. J. P. Williams, eds.), Springer-Verlag, Berlin, pp. 49–81.

Joshi, N. B., and Shamoo, A. E., 1987, *Biophys. J.* **51**:185–191.

Kupke, D. W., and Fox, J. W., 1989, *Biochemistry* **28**:4409–4415.

La Mar, G. N., 1973, in *NMR of Pramagnetic Molecules* (G. N. La Mar, W. D. Horrocks, Jr., and R. H. Holm, eds.), Academic Press, New York, pp. 85–126.

La Mar, G. N., Horrocks, W. D., Jr., and Holm, R. H. (eds.), 1973, *NMR of Paramagnetic Molecules,* Academic Press, New York.

Luk, C. K., 1971, *Biochemistry* **10**:2838–2843.

Madras, J. S., and Brittain, H. G., 1980, *Inorg. Chem.* **19**:3841.

Martin, R. B., and Richardson, F. S., 1979, *Q. Rev. Biophys.* **12**:181–209.

Matthews, B. W., and Weaver, L. H., 1974, *Biochemistry* **13**:1719–1725.

Meares, C. F., and Rice, L. S., 1981, *Biochemistry* **20**:610–617.

Meares, C. F., and Wensel, T. G., 1984, *Acc. Chem. Res.* **17**:202–209.

Meares, C. F., Yeh, S. M., and Stryer, L., 1981, *J. Am. Chem. Soc.* **103**:1607–1609.

Mikkelsen, R. B., and Wallach, D. F. H., 1974, *Biochim. Biophys. Acta* **363**:211–218.

Morley, J. P., Saxe, J. D., and Richardson, F. S., 1982, *Mol. Phys.* **379**:407.

Muetterties, E. L., and Wright, C. W., 1967, *Q. Rev. Chem. Soc.* **21**:109.

Mulqueen, P., Tingey, J. M., and Horrocks, W. D., Jr., 1985, *Biochemistry* **24**:6639–6645.

Navon, S., Stavola, M., and Skeats, M. G., 1981, *J. Inorg. Nucl. Chem.* **43**:575–578.

Nieboer, E., 1975, in *Structure and Bonding,* Vol. 22 (J. D. Dunitz, P. Hemmerich, R. H. Holm, J. A. Ibers, C. K. Jørgensen, J. B. Neilands, D. Reinen, and R. J. P. Williams, eds.), Springer-Verlag, Berlin, pp. 1–47.

Patil, G. S., Dorman, N. J., and Cornwell, D. G., 1979, *J. Lipid Res.* **20**:663–668.

Peacock, R. D., 1975, in *Structure and Bonding,* Vol. 22 (J. D. Dunitz, P. Hemmerich, R. H. Holm, J. A. Ibers, C. K. Jørgensen, J. B. Neilands, D. Reinen, and R. J. P. Williams, eds.), Springer-Verlag, Berlin, pp. 83–122.

Petersheim, M., and Sun, J., 1989, *Biophys. J.* **55**:631–636.

Petersheim, M., Blodnieks, J., and Halladay, H. N., 1989a, in *Biological and Synthetic Membranes* (A. R. Butterfield, ed.), Liss, New York, pp. 87–96.

Petersheim, M., Halladay, H. N., and Blodnieks, J., 1989b, *Biophys. J.* **56**:551–557.

Reid, M. F., and Richardson, F. S., 1985, *J. Phys. Chem.* **88**:3579–3586.

Reisfeld, R., 1975, in *Structure and Bonding,* Vol. 22 (J. D. Dunitz, P. Hemmerich, R. H. Holm, J. A. Ibers, C. K. Jørgensen, J. B. Neilands, D. Reinen, and R. J. P. Williams, eds.), Springer-Verlag, Berlin, pp. 123–175.

Reisfeld, R., 1976, in *Structure and Bonding,* Vol. 30 (J. D. Dunitz, P. Hemmerich, R. H. Holm, J. A. Ibers, C. K. Jørgensen, J. B. Neilands, D. Reinen, and R. J. P. Williams, eds.), Springer-Verlag, Berlin, pp. 65–97.

Reuben, J., 1979, in *Handbook on the Physics and Chemistry of Rare Earths,* Vol. 4 (K. A. Gschneidner, Jr., and L. R. Eyring, eds.), North-Holland, Amsterdam, pp. 515–552.

Reuben, J., and Elgavish, G. A., 1979, in *Handbook on the Physics and Chemistry of Rare Earths,* Vol. 4 (K. A. Gschneidner, Jr., and L. R. Eyring, eds.), North-Holland, Amsterdam, pp. 483–514.

Rhee, M.-J., Sudnick, D. R., Arkle, V. K., and Horrocks, W. D., Jr., 1981, *Biochemistry* **20**:3328–3334.

Richardson, F. S., 1980, *Inorg. Chem.* **19**:2806.

Richardson, F. S., 1982, *Chem. Rev.* **82**:541–552.

Ryan, J. L., and Jørgensen, C. K., 1966, *J. Phys. Chem.* **70**:2845.

Sen, A. C., Chowdhury, M., and Schwartz, R. W., 1981, *J. Chem. Soc. Faraday Trans. II* **77**:1293.

Sinha, S. P., 1976a, in *Structure and Bonding,* Vol. 25 (J. D. Dunitz, P. Hemmerich, R. H. Holm, J. A. Ibers, C. K. Jørgensen, J. B. Neilands, D. Reinen, and R. J. P. Williams, eds.), Springer-Verlag, Berlin, pp. 69–149.

Sinha, S. P., 1976b, in *Structure and Bonding,* Vol. 30 (J. D. Dunitz, P. Hemmerich, R. H. Holm, J. A. Ibers, C. K. Jørgensen, J. B. Neilands, D. Reinen, and R. J. P. Williams, eds.), Springer-Verlag, Berlin, pp. 1–64.

Snyder, A. P., Sudnick, D. R., Arkle, R. K., and Horrocks, W. D., Jr., 1981, *Biochemistry* **20**:3334–3339.

Sowadsky, J., Cornick, G., and Kretsinger, R. H., 1978, *J. Mol. Biol.* **124**:123–132.

Stevens, B. R., and Kneer, C., 1988, *Biochim. Biophys. Acta* **942**:205–208.

Sun, J., 1990, Master's thesis, Seton Hall University, South Orange, N.J.

Sun, J., and Petersheim, M., 1990, *Biochim. Biophys. Acta* **1024**:159–166.

Svetashaev, A. G., and Tsvirko, M. P., 1984, *Teor. Eksp. Khim.* **20**:696–701.

Thomas, D. D., Carlsen, W. F., and Stryer, L., 1978, *Proc. Natl. Acad. Sci. USA* **75**:5746–5760.

Thompson, L. C., 1979, in *Handbook on the Physics and Chemistry of Rare Earths,* Vol. 3 (K. A. Gschneidner, Jr., and L. R. Eyring, eds.), North-Holland, Amsterdam, pp. 209–297.

Wybourne, B. G., 1965, *Spectroscopic Properties of Rare Earths,* Wiley, New York.

Yan, F., Copeland, R. A., and Brittain, H. G., 1982, *Inorg. Chem.* **21**:1180.

Zolin, V. F., Koreneva, L. G., and Tsaryuk, V. I., 1975, *Biofizika* **20**:194.

Zuxuan, Z., Yuanzhen, Z., Xinguang, W., Rongsan, C., and Zhenzhang, R., 1985, *J. Less-Common Met.* **112**:401–409.

Chapter 3

Continuous, On-Line, Real Time DNA Sequencing Using Multifluorescently Tagged Primers

John Brumbaugh, Lyle Middendorf, Dan Grone, and Jerry Ruth

1. INTRODUCTION

DNA sequencing is a relatively new technique, just over a decade old. Scientifically, it is a very important technique because the DNA sequence gives us the ultimate resolution of the genetic material to the single base level and thus its complete information.

Abbreviations used in this chapter: DMSO, dimethyl sulfoxide; dATP, deoxyadenosine triphosphate; ddATP, dideoxyadenosine triphosphate; dCTP, deoxycytidine triphosphate; ddCTP, dideoxycytidine triphosphate; dGTP, deoxyguanosine triphosphate; ddGTP, dideoxyguanosine triphosphate; dNTP, deoxynucleotide triphosphate; ddNTP, dideoxynucleotide triphosphate; dTTP, deoxythymidine triphosphate; ddTTP, dideoxythymidine triphosphate; EDTA, ethylenediaminetetraacetic acid; EMBL, European Molecular Biology Laboratory; FITC, fluorescein isothiocyanate; GCG, (University of Wisconsin) Genetics Computer Group; HPLC, high-pressure liquid chromatography; O.D., optical density; PAGE, polyacrylamide gel electrophoresis; PEG, polyethylene glycol; PMT, photomultiplier tube; RPHPLC, reverse-phase high-pressure liquid chromatography; TEMED, tetramethylethylenediamine; TES, Tris–EDTA salt solution.

JOHN BRUMBAUGH • School of Biological Sciences, University of Nebraska, Lincoln, Nebraska 68588.   LYLE MIDDENDORF and DAN GRONE • Li-Cor, Inc., Lincoln, Nebraska 68504.   JERRY RUTH • Molecular Biosystems, Inc., San Diego, California 92121.

1.1. Original Methods

Originally, DNA was sequenced using one of two methods. Maxam and Gilbert (1977, 1980) devised a method that chemically degrades DNA selectively between specific bases. Sanger *et al.* (1977) developed an enzymatic method based on the use of chain-terminating dideoxynucleotides.

Both methods produced four groups of radioactively labeled fragments specific for each base type (adenine, A; thymine, T; guanine, G; cytosine, C). The lengths of the fragments in each group corresponded to cleavage (Maxam and Gilbert, 1977, 1980) or termination (Sanger *et al.,* 1977) at each specific base. Electrophoresis was then used to separate the strands in a gel according to their lengths. After autoradiography of the electrophoresed sample(s), the relative positions of the strands were indicated by dark bands on the autoradiogram. The sequence information was then determined by "reading" the bands beginning at the bottom of the autoradiogram (shortest strands) and proceeding upward.

These classical methods of DNA sequencing have inherent limitations and drawbacks. Autoradiography is a static detection method. Therefore, spatial parameters limit the ability to read sequence. As DNA strands increase in length, the band-to-band intervals become so small that sequence cannot be read at the top of the autoradiogram (gel) because they remain together as one band.

The average length of DNA sequence that can be determined with a single reaction in a single load is approximately 300 bases (Wada, 1987). This can be increased by using longer gels (Ansorge and Barker, 1984) or gradient gels (Ansorge and Labeit, 1984; Olsson *et al.,* 1984; Biggen *et al.,* 1983) which provide more fragment separation. This makes gel manufacture more difficult, however. Alternatively, longer sequences can be determined from a single reaction by sequencing the overlaps from multiple loadings of aliquots of the same sample reaction loaded at various time intervals after the commencement of electrophoresis (BRL, 1980). This, in essence, mimics a long gel by allowing more time for fragment separation in the earlier loads.

Autoradiography also precludes that sequencing be done by the batch method. Electrophoresis is carried out until bands are spread over the length of the gel. The glass plates are separated, the gel carefully removed and exposed to X-ray film. The gel may be dried before exposure to the film. After the exposure period, which is at least overnight and may be as long as 2 or 3 days, the film is developed and the sequence determined.

In order to generate large amounts of data with the batch method, an assembly line must be developed. The assembly line must have high-quality control so that the data produced from batch to batch can be compared. The individuals doing the repetitive steps must have excellent manual dexterity and be well trained. This requirement for skilled workers at every step makes sequencing very costly.

The use of radioactivity also adds complicating factors. There is the cost and also the limited shelf life due to the half-life of the radioisotope used. Personnel who handle radioactive materials must be monitored and the laboratory designed to reduce hazards and provide for safe disposal. Thus, these original methods, which are costly and time-consuming, place limitations upon the amount of data that can be produced on large-scale sequencing projects.

Efforts to reduce some of these limitations have been successfully applied by several laboratories. A group in Japan has developed precast electrophoretic gels (Wada, 1984). These gels are mass-produced, disposable, have a uniform quality, and can be stored. They are used like cassettes in the sequencing operation.

Other laboratories have a large number of electrophoretic apparatuses which use gels cast between very large glass plates (20 cm × 100 cm) (Garoff and Ansorge, 1981; Ansorge and Barker, 1984).

Computer programs, suitable for automatically reading autoradiographic films, have been developed (Elder *et al.,* 1986; West, 1988). The software and hardware are now commercially available.

Multiplex sequencing (Church and Gilbert, 1984; Church and Kieffer-Higgins, 1988) allows many sequences to be determined from a single set of reactions and a single electrophoretic separation. The probes or primers are different for each sample and the DNA is transferred and affixed to membranes where it is multiply probed, one at a time, with each probe or primer and autoradiographed.

1.2. The Need for Automation

To date, specific genes (usually expressed genes) from various organisms have been sequenced. The choices were made by individual researchers interested in solving particular research questions. More recently, it has become evident that much could be learned from determining the entire DNA sequence of a given organism (National Research Council, 1988). The organism that has come to the forefront is man.

The haploid genome of humans contains approximately 3 *billion* base pairs. The DNA is packaged as 23 different kinds of chromosomes of varying lengths of DNA. (The number is 24 when the male-specific Y chromosome is included.) Each of these chromosomes has specialized DNA sequences at its ends (telomeres) and in the regions that attach to the spindle fibers during cell division (centromeres). Determining the sequence of DNA in a given chromosome from telomere to centromere to telomere and understanding the arrangements and sequences of the genes between these regions and the sequences of DNA between genes would tell us much about cell division and the regulation of gene activity. A complete DNA sequence would also be an invaluable resource in the study of genetic disease.

Some important experimental organisms such as mice, *Drosophila, C. elegans,* and *Arabidopsis* will also be sequenced. Agriculturally important plant and animal species such as maize, wheat, poultry, cattle, and swine also need to be sequenced if we expect to increase food production on a worldwide basis.

In order to complete this enormous task, DNA sequencing must become rapid and relatively inexpensive. Thus, laboratories worldwide have begun to develop ways of automating DNA sequencing.

1.3. Review of Automated DNA Sequencing Technologies

The reported automated DNA sequencing techniques developed thus far have used the Sanger dideoxy method (1977). The authors are quite certain that automated techniques using the Maxam–Gilbert (1977, 1980) method are also being developed.

The first type of automated DNA sequencing was reported by Beck and Pohl (1984). In this method, a short electrophoretic gel was used to separate the DNA fragments. A membrane was moved along the bottom of the sequencing gel so that the DNA bands were transferred to the membrane as they reached the end of the gel. A very long membrane was produced which contained the DNA bands. This technique did not eliminate batch processing or autoradiography of the membrane. Later, a nonradioactive method was developed (Beck, 1987) which used a biotinylated nucleotide analogue which allowed strepavidin binding. The strepavidin was conjugated to an enzyme which produced a colored end product in the presence of the designated substrate. Batch processing of the membranes was still required, however.

The first form of continuous, on-line DNA sequencing was reported by Smith *et al.* (1986). This method uses four different primers, each conjugated to a different fluorophore. Each dideoxy reaction was carried out (without radioactivity) with a different primer so a given base type was associated with that fluorophore. The four reaction mixtures were then electrophoresed in the same channel. Using a laser, the fluorophores were excited and the resulting signals collected with a photomultiplier tube at a fixed distance from the top of the gel, thus producing data as the DNA bands passed the detector. This method has the advantage in that there are no between-lane migration differences since all the bands are in a single lane. The limitations of this technique revolve around the complexities related to exciting, collecting, and analyzing data generated by four fluorophores of different molecular weights and different excitation and emitting wavelengths.

A second continuous, on-line detection method was developed by Ansorge *et al.* (1986, 1987). A single fluorescent dye was attached to a single primer which is used in the four dideoxy reactions (without radioactivity). In this case, the laser beam was focused across the gel between the glass plates. The data are collected by a fixed detection system. This method has an advantage in that the gel does not have to be mechanically scanned. It has the disadvantage that the gel temperature must be controlled to minimize band migration differences since a four-lane format is needed to determine sequence. In addition, the larger spot size associated with a parallel laser beam may limit resolution of longer strands. Because a single fluorophore is used, the detection method is much simpler and a small amount of DNA can be detected.

Both the four-dye, single-lane method developed by Smith *et al.* (1986) and the single-dye, four-lane method developed by Ansorge *et al.* (1986, 1987) depend on fluorescently tagged primers to produce the sequencing data, as does the method described in this chapter. A third version of continuous, on-line DNA sequencing was developed by Prober *et al.* (1987). In this case, four different fluorescently tagged dideoxy analogues were used (one for each base) which were terminally incorporated into the DNA strands. Each analogue contained a fluorophore whose emitting wavelength differed slightly from the other three. Computer analysis of the ratio of the various wavelengths collected was used to determine which type of band was passing the detector at a particular time.

Kambara *et al.* (1988) reported a variation of the technique described by Ansorge *et al.* (1986, 1987) of exciting the bands by introducing the laser excitation laterally between the glass plates. They tried two different primers. One was similar to that of Smith *et al.* (1986) and Ansorge *et al.* (1986, 1987) with the fluorophore attached at the 5' end of the molecule. The second type of primer had the fluorescent complex attached to a base internal to the ends of the molecule.

Other groups have developed continuous, on-line DNA sequencers which detect radioactivity rather than fluorescence through the use of fixed scintillation detectors (Nagai *et al.*, 1987; Page, 1988). The concepts are the same as those used for continuous, on-line sequencing with fluorescent markers.

The method described in this chapter is a combination of the fluorescence techniques previously described (Brumbaugh *et al.*, 1988; Middendorf *et al.*, 1988). The technology has been operational on a routine basis since mid-1986. Like Ansorge *et al.* (1986, 1987), a single fluorophore is used for all reaction lanes. Like Smith *et al.* (1986) and Prober *et al.* (1987), the laser beam is scanned back and forth across the gel rather than being introduced between the two glass plates. The unique features are: (1) two fluorescent molecules are attached to each primer which increases the signal and (2) the bands are imaged in two dimensions rather than as simple curves so that the relationships between bands in a sample can be accurately determined.

Dideoxy protocols for this method have been determined that provide sequence for 600 bases with typically less than 1% error in a single load. Electrophoresis is carried out at 20–22 W and the bands are detected 24 cm from the bottom of the loading wells with a scanning fluorescence detector. Bands are imaged on a TV screen in two dimensions. The sequences can be read from the TV screen manually or semiautomatically by using a simple software program. The system allows more bases to be read with a lower error rate than any other reported automated sequencing method.

2. MATERIALS AND METHODS

2.1. Sample Preparation

DNA is prepared for automated sequencing by a modest scale up of minipreparations (Messing, 1983). More than the minimum amount of DNA is prepared so that multiple reactions and runs of the same sample can be performed without the additional time-consuming step of isolation. It is not necessary to purify the DNA with cesium chloride ultracentrifugation or separation columns. The quality of DNA produced by simple isolations is adequate for consistent sequencing.

Single-stranded DNA is prepared for sequencing from two different biological systems. DNA fragments that are cloned into the single-stranded DNA bacteriophage, M13 (Messing, 1983), are simply isolated from cultures of M13. The second method uses plasmids that contain the M13 origin of replication such as pBluescript M13 (Ausubel *et al.,* 1987). Fragments to be sequenced are cloned into these plasmids which produce single-stranded DNA phages when infected with an M13 helper virus. The method for preparing DNA from M13 will be described first, followed by the modifications used with the helper virus/plasmid systems.

E. coli strain JM109 is infected with M13. JM109 is the preferred host because it is both rec A⁻ and r⁻, which eliminates recombination and restriction enzyme modification, respectively (Ausubel *et al.,* 1987). Stationary cultures of JM109 are produced by placing one colony into 20 ml of M9 glucose minimal medium and shaking overnight in a 37°C waterbath (Messing, 1983). The JM109 colonies are maintained on M9 plates (Messing,

1983). M9 medium is essential in maintaining the desired JM109 genotype. Flasks containing 20 ml of the enriched medium 2xYT (Messing, 1983) are prepared for each separate viral stock that is to be propagated. Stocks used for sequencing should have a titer of $>10^{10}$ pfu/ml. Two hundred microliters of the overnight JM109 culture is placed in each flask. These flasks are then shaker incubated at 37°C for 1 hr. After this time, 200 μl of each of the viral stocks is pipetted into the flasks which are appropriately coded. These flasks are then shaker incubated overnight with vigorous agitation. The following morning, a 1-ml sample from each flask is taken and saved for backup purposes in case the DNA sample is lost. Each sample is then spun at 7700g for 10 min. The supernatant which contains the phages is transferred to clean tubes which contain 4 ml of a 20% polyethylene glycol (PEG; 8000 daltons) 2.5 M NaCl solution. The tubes are incubated for 40 min at room temperature to precipitate the viruses. After precipitation, the tubes are centrifuged for 15 min at 17,400g. The supernatant is carefully removed with a pipette and the inside walls of the centrifuge tubes are wiped with a tissue to remove any remaining PEG solution.

The pellet in each tube is resuspended with 2.1 ml of TES (20 mM Tris-HCl, 10 mM NaCl, 0.1 mM Na_2 EDTA, pH 7.5) and divided into three 1.5-ml microfuge tubes (700 μl each). The microfuge tubes are color-coded to keep track of samples throughout the procedure. To each 700 μl of viral suspension is added 700 μl of saturated phenol. The microfuge tubes are vortexed, left to sit 5 min, vortexed again, and then microfuged briefly. The upper layer is transferred to clean tubes and two additional extractions performed using a 1:1 mixture of phenol–chloroform/isoamyl alcohol (24:1) and chloroform/isoamyl alcohol.

The DNA is precipitated by adding 30 μl of sodium acetate (3 M, pH 5) and 700 μl of 95% chilled ethanol. The microfuge tubes are inverted once, chilled on ice for 10 min, and then centrifuged at 4°C for 30 min. The supernatant is discarded and the pellet is rinsed with isopropanol and again centrifuged at 4°C for 10 min. The supernatant is then discarded and the tubes left to dry for at least 1 hr. Frequently the tubes are left to dry overnight. The resulting DNA pellet is suspended in 16 μl of TES. The resuspension is done serially so that a single sample of three tubes ends up being suspended in 16 μl of TES. It is desirable to obtain a concentration of DNA > 0.5 μg/μl. These preparations usually produce DNA concentrations in the 1–2 μg/μl range. DNA concentrations are determined by measuring absorbance at 260 nm and calculating the concentration of DNA using 1 O.D. $\cong 40$ μg/ml.

The helper virus/plasmid system is preferred because it produces very consistent results with higher concentrations of DNA. This is due to the fact that kanamycin resistance is incorporated into the helper virus (Vieira and

Messing, 1987). Kanamycin selection allows only virally infected bacteria to replicate. With the M13 system, uninfected bacteria can grow, thus reducing the yield of viral particles.

When the plasmid preparations are used, they can be provided as stabs, slants, or streak plates. Fresh streak plates are prepared from these stock cultures and several colonies placed into 20 ml of 2xYT containing 20 μl of ampicillin stock solution (150 mg/ml) for a final concentration of 150 μg/ml. Ampicillin selects for the plasmids which contain the ampicillin resistance gene. Each culture is shaker incubated at 37°C for 4 hr. After this incubation, 10 μl of stock helper virus, M13K07 (titer > 10^{11} pfu/ml), is added. The cultures are then shaker incubated for an additional hour and the O.D. 660 determined. Each culture is then diluted with 2xYT to an O.D. 660 $\cong$ 0.1 and 20 μl of kanamycin stock solution (70 mg/ml) added for a final concentration of 70 μg/ml. The cultures are then shaker incubated overnight at 37°C. The resulting culture contains two genetically different phages. One is the helper virus, and the other is one strand of the plasmid packaged in the viral protein jacket. The helper virus DNA does not interfere with sequencing because it does not contain the sequences to which the sequencing primers hybridize. From this point on, the isolation procedure is the same as that for M13.

2.2. Synthesis of Fluorescently Labeled Primers

A deoxyuridine analogue with a primary amine "linker arm" of 12 atoms attached at C-5 was synthesized as previously published (Ruth *et al.*, 1985; Jablonski *et al.*, 1986). Synthesis of the analogue consists of derivatizing 2′-deoxyuridine through organometallic intermediates to give 5-(methyl propenoyl)-2′-deoxyuridine. Reaction with dimethoxytritylchloride produces the corresponding 5′-dimethoxytrityl adduct. The methyl ester is hydrolyzed, activated, and reacted with an appropriately monoacylated alkyl diamine. After purification, the resultant "linker arm" nucleosides are converted to nucleoside analogues suitable for chemical oligonucleotide synthesis. The structure of the "linker arm" analogue is shown in Fig. 1.

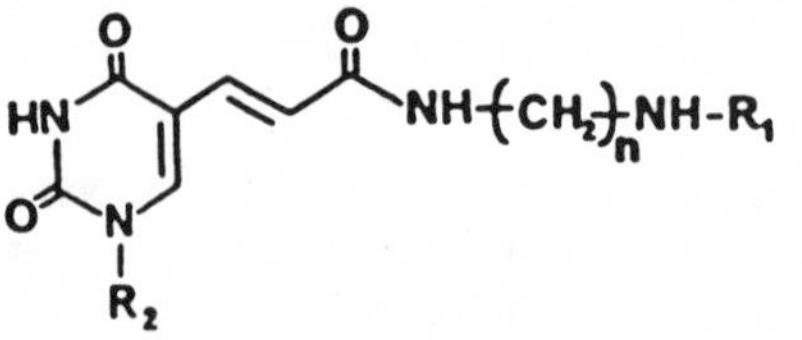

FIGURE 1. Chemical structure of linker arm analogue. n-7; R_1, fluorescein isothiocyanate (FITC); R_2, deoxyribose. Reprinted from *Proc. Natl. Acad. Sci. USA* **85**:5610–5614 (1988).

A 19-base M13 primer (5'-dGGTTTTCCCAGTCACGACG-3') was made which included two "linker arm" bases (T) at positions 5 and 12. The primer was synthesized using modified phosphorarmidite chemistry and purified to electrophoretic and chromatographic homogeneity by reverse-phase HPLC (RPHPLC) (Ruth, 1984).

To a solution of 50 nmoles of 19mer "linker arm" oligonucleotide in 25 μl of 500 mM sodium bicarbonate (pH 9.4) was added 20 μl of 300 mM fluorescein isothiocyanate (FITC) in dimethyl sulfoxide (DMSO). The mixture was agitated at room temperature for 6 hr. The oligonucleotide was separated from free FITC by elution from a 1 × 30-cm Sephadex G-25 column with 20 mM ammonium acetate (pH 6), combining fractions in the first UV-absorbing peak. Analysis by analytical 20% PAGE indicated the reaction was complete, with fluorescent oligomer electrophoresing slower than nonfluoresceinated oligomer by the equivalent of 1 nucleotide unit. The FITC-oligomer was purified by preparative RPHPLC using an 8.3-cm Perkin–Elmer Pecosphere C-8 (3 μm) column eluted with a linear gradient of 7–35 vol% acetonitrile in 100 mM triethylammonium acetate (pH 7.0) over 20 min at 1.0 ml/min. Analyses were by a Waters model 490 multiple wavelength absorbance detector with simultaneous detection at 260 and 495 nm. The product was concentrated and ethanol precipitated to recover 33 nmoles (65%) of fluoresceinated oligonucleotide. Products were homogeneous by PAGE and RPHPLC, kinased normally, and had A260/490 ratios of 2.08 as predicted for such oligomer–fluorophore conjugates at pH 8. The product was used directly for hybridization.

Other primers have been synthesized and successfully used in addition to the M13 forward primer. These include an M13 reverse primer and primers in the T3 and T7 regions of pBluescript.

2.3. Activity of Fluorescently Labeled Primers

For fluorescent sequencing to be reliable, the fluorescent dyes attached to the primers must not interfere with their biological activity. Primer, containing two "linker arms" as described in the previous section, but with no fluorescent groups attached, was used in sequencing reactions using standard dideoxy protocols with [^{32}P]dCTP (Sanger *et al.*, 1977). The test specimen was a 1-kilobase (kb) portion of pBR325 cloned into M13 mp18 at the *Hind*III site and terminating at the *Eco*RI site (Prentki *et al.*, 1981). Similar reactions were done using difluoresceinated primer with FITC attached to each linker arm.

The autoradiograms showed that neither the linker arms nor the linker arms with FITC attached interfered with biological activity and allowed conventional sequence data to be generated (Fig. 2). As expected, the bands

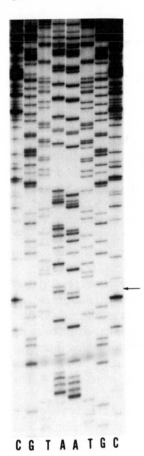

FIGURE 2. Autoradiogram of sequence of pBR325 inserts using nonfluoresceinated primer (right four lanes) and difluoresceinated primer (left four lanes). The arrow points to the *Hind*III site (AAGCTT). The difluoresceinated primer produces readable bands with all four lanes migrating about one base slower than the bands produced using the nonfluoresceinated primer. Reprinted from *Am. Biotechnol. Lab.* **6**(6):14–22 (1988).

C G T A A T G C

containing the difluoresceinated primer migrated more slowly in the gel than the bands produced using the linker-arm-only primer but still yielded a normal pattern (Fig. 2). The difluoresceinated primer therefore served as a reliable substrate for Klenow polymerase extension and thus was suitable for fluorescent DNA sequencing. The presence of the two fluoresceins did not seem in any way to interfere with hybridization, extension, or the production of normal sequencing data.

2.4. Dideoxy Reactions for Automated Sequencing

The DNA isolated as previously described is used as template DNA for the sequencing reactions. Five micrograms of template DNA (approximately 2 pmoles) and 50 ng of difluoresceinated primer (approximately 7.2 pmoles) are hybridized in 12.5 μl of polymerase buffer. Hybridization is effected by heating the template/primer mixture to 90°C and allowing it to cool slowly for 45 min. Sometimes the DNA is heated to 90°C and quickly chilled on ice to reduce the formation of secondary structure and then hybridized with a primer at a specific hybridization temperature, usually 60°C for 5 min.

After hybridization, 1 μl of 0.1 M dithiothreitol, 15 units of the Klenow fragment of DNA polymerase I, and enough H_2O to bring the final volume to 19 μl are added and mixed.

To 3 μl of the aforementioned template/primer/Klenow mixture is added 1 μl of deoxy nucleotide triphosphate (dNTP) mix and 1 μl of dideoxynucleotide triphosphate (ddNTP) mix for each respective base type, giving four reaction tubes. The final concentration of the dNTPs is 65.6 μM except the reaction-specific deoxynucleotide concentration which was 6.45 μM. The final dideoxy concentrations are as follows: ddATP, 400 μM; ddTTP, 600 μM; ddGTP, 100 μM; ddCTP, 100 μM. The 7-deaza-2-deoxy dGTP analogue is used in place of dGTP to reduce GC-rich compressions (Mizusawa et al., 1986; Barr et al., 1986).

Reaction-specific mixtures of dNTPs are prepared from 10 mM stock solutions which are kept frozen at $-100°C$. Each of these stocks is diluted just before each reaction to 1 mM by using 7 μl of the dNTP plus 63 μl of H_2O and stored on ice. Two microliters of the 1 mM dATP plus 20 μl of each of the 1 mM solutions of dTTP, dGTP, and dCTP are mixed together. This is the A reaction dNTP mixture. The T, G, and C dNTP mixes are made in the same way except that for the T reaction, only 2 μl of dTTP is added and 20 μl of dATP is used. For the G reaction, only 2 μl of dGTP is added to 20 μl of each of the other dNTPs. For the C reaction, only 2 μl of dCTP is added to 20 μl of each of the other dNTPs.

The ddNTPs are also made up as 10 mM stock solutions which are kept frozen at $-100°C$. The mixtures are made according to specific formulas. For ddATP, 4 μl of stock solution is added to 16 μl of H_2O. For ddTTP, 6 μl of stock solution is added to 14 μl of H_2O. For ddGTP and ddCTP, 4 μl of stock solution is added to 76 μl of water.

Each reaction tube is incubated at 30°C for 40 min and the reactions stopped by adding 5 μl of stop buffer. The stop buffer consists of 0.1% (w/v) xylene cyanol, 0.1% (w/v) bromphenol blue, 10 mM Na_2 EDTA, 95% (v/v) deionized formamide. This procedure is similar to standard dideoxy sequencing protocol (Sanger et al., 1977) except that the amounts of template, primer, Klenow fragment, and nucleotides are increased. No radioactivity is used. Reaction mixtures with stop buffer added can be frozen at $-20°C$ and stored for a week without appreciable loss in sequencing quality.

This reaction protocol was standard for use with the Klenow fragment of DNA polymerase I and for samples whose A+T/G+C ratios varied between 0.67 and 1.50. The reactions produce an array of readable bands of up to 600 bases and sometimes even further. If the sample is AT-rich (A+T/G+C > 1.50), then it is necessary to increase the dATP and dTTP pools to allow for bands of the longer DNA fragments to be produced. When the samples are GC-rich (A+T/G+C < 0.67), then it is necessary to increase the dGTP and dCTP pools. Remember that the dGTP pool is actually the 7-deaza analogue of GTP which significantly eliminates most GC compressions. Sometimes GC compressions remain a problem, particularly when the A+T/G+C ratio is <0.50. Such compressions can be eliminated by using Taq polymerase (New England Biolabs), and carrying out the reactions at 60°C and running the gels at 60°C. This successfully eliminates the secondary structures in the GC-rich DNA.

Standard sequencing protocols with some modest scaleup are sufficient to produce very consistent and reliable sequencing data using fluorescence detection.

2.5. Gel Electrophoresis

For automated sequencing, gel electrophoresis is routine. The gels are shorter because the data are collected near the bottom of the gels in a continuous manner. Band separation is effected by the distance traveled through the gel rather than needing to be spatially separated along the length of the gel. Shorter gels are obviously one of the advantages of using continuous, on-line detection.

Six percent acrylamide gels (5% bis-acrylamide) with 8 M urea are routinely used. The gel thickness is 0.35 mm. The buffer is 133 mM Tris base, 44 mM boric acid, and 2.8 mM EDTA with the pH adjusted to 8.9 with NaOH at 50°C to minimize the reduction in fluorescence that occurs at lower pH values.

The gels are 18 cm × 30.5 cm and poured between two borosilicate glass plates (Schott, Tempax) 3.2 mm thick. Soda lime glass (3.2 mm thick, window glass) can also be used. Sixteen-well combs are used to produce wells 3.4 mm × 4.8 mm which allows four samples of four lanes each to be run simultaneously. Prior to gel manufacture, the gel side of the plate next to the detector is treated with Sigmacote (Sigma). The top 4 cm of the gel side of the other plate is treated with a binding silane, γ-methacryloxypropyltrimethoxysilane (Garoff and Ansorge, 1981).

The liquid gel solution is filtered through a 0.22-μm filter cup by vacuum to degas the solution and eliminate particulate matter. Then TEMED

(0.02%) and 10% ammonium persulfate (0.6%) are added to hasten polymerization and the gels immediately poured. After pouring, the gels are allowed to harden for at least 1 hr.

After hardening, the gels are prerun at 1200 V for 0.5–1.0 hr. At this point, the urea is flushed out of each well and the samples loaded. Samples are generally loaded in the ATGC format. Between 1.3 and 1.6 μl of sample is loaded in each lane. This is equivalent to 100–125 ng (40–50 fmoles) of template DNA per well or 400–500 ng (160–200 fmoles) of template per sample. Immediately after loading, 2000 V is applied for 5 min to drive the samples quickly into the gel. The power supply is set to run at a constant power of 20–22 W (approximately 40 V/cm). A thermostatically controlled temperature plate placed against one side of the glass gel sandwich is regulated at 50°C during a standard run. The temperature plate has a variable temperature control so other temperatures can be used. These standard conditions produce data at an average of 1–1.5 bases per minute per sample.

Some variables have been used. For example, 4% acrylamide gels produce data at a more rapid rate but problems with well morphology are encountered. It has been determined (Ansorge and Barker, 1984; Kambara *et al.*, 1988) that 4% gels provide band separation over the widest range of fragment lengths. Methods for producing 4% gels with good well morphology are being developed. As previously mentioned, sometimes the gel temperature was regulated to 60°C to eliminate secondary structure in the migrating DNA fragments.

2.6. Detection and Imaging

The detection and imaging system consists of an argon laser which is used to excite the sample whose emissions are collected by a moving fluorescent microscope. The emissions produced by laser excitation are filtered and collected with a photomultiplier tube (PMT). The collected signals are converted to digital information by an analog/digital converter and then processed by a computer. The computer is programmed to display the data in *two* dimensions on a video screen. The horizontal components of the image are produced by the back-and-forth movement of the fluorescent microscope while the vertical components of the image are produced by the movement of the DNA fragments past the detector due to electrophoresis. A block diagram of the detection and imaging system is shown in Fig. 3.

A 10-mW argon laser (model 2001-10SL, Cyonics, Sunnyvale, Calif.) is used as the excitation source. FITC which is attached to the linker arms of the primer has an absorption peak at 490 nm. The 488-nm laser output radiation is therefore used to excite the FITC. The laser output is filtered by a

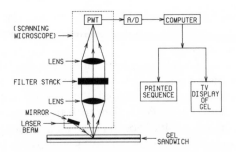

FIGURE 3. Block diagram of detection and imaging system. PMT, photomultiplier tube; A/D, analog-to-digital converter. Reprinted from *Proc. Natl. Acad. Sci. USA* **85**:5610–5614 (1988).

1-nm band-pass filter centered at 488 nm (Corion, D1-488-F). The filter is used to eliminate stray 514.5-nm radiation that also comes from the argon laser.

The excitation radiation produced by the laser is chopped at 10 kHz by a rotating chopping wheel (5000 rpm, 128 slots per revolution). The laser signal is chopped to reduce the inherent background fluorescence of the glass and gel. Since FITC has a short time constant compared to the long time constant of the background fluorescence, the bands can be readily distinguished by chopping.

The excitation radiation is reflected by a mirror mounted on the movable microscope stage so that it enters the glass/gel sandwich at Brewster's angle (56°) which minimizes reflected stray light and maximizes the amount of energy imparted to the sample. The laser beam is oriented so that the light vibrations are parallel to the plane of incidence. After being reflected by the mirror but before entering the glass/gel sandwich, the excitation radiation is focused to a spot diameter of $\cong 30$ μm by a 25.4-mm-focal-length lens (Melles Griot 01 LPX 037) located on the movable microscope stage.

The emitted radiation is collected by the microscope that is scanned back and forth across the gel sandwich 24 cm from the bottom of the loading wells. The relatively rapid scan rate (5 cm/sec) minimizes fluorescent quenching that would occur if the laser spot were to remain in one position for a relatively long period of time. Scanning is accomplished by a stepper motor-controlled lead screw.

The objective lens is a 20-mm-focal-length aspheric lens f#/1 (Rolyn optics #17.1055) that focuses the emitted radiation at infinity. The radiation is then filtered by a 20-nm band-pass filter centered at 520 nm (Corion S10-520-F), two sharp-cut orange filter glasses (Schott OG515, 3 mm thick), and a 550-nm short-pass filter (Corion LS-55-F) that filters out the Raman background (585 nm) of the water (3400/cm Raman shift) in the gel. After filtering, the radiation is focused by a 100-mm-focal-length lens (Melles Griot 01 LPX 177) through a spatial filter onto a PMT (Hamamatsu R928). The PMT is operated at a voltage of between -800 and -950 V.

The electrophoretic apparatus and optical scanner are housed in a light-proof closure to minimize stray room light as well as provide protection to the user against laser light and high-voltage supplies. When the door of the electrophoretic housing is opened, the laser and high-voltage supplies are shut down. A float is located in the upper buffer tank so that if a leak occurs and buffer level falls, the shutdown procedure is automatically initiated. In addition, electrophoresis current is monitored for open circuit conditions.

The PMT output current is amplified by a current amplifier (PAR model 181). A lock-in amplifier (PAR model 5101) is used to measure only the 10-kHz component of the PMT signal. A reference signal for the lock-in amplifier is generated by the chopper wheel. The signal is then filtered by an eight-pole Bessel filter (Frequency Devices model 902LPF1) having a cutoff of 10 Hz and then digitized at 256 conversions per second giving 512 conversions for a 2-sec scan (10 cm).

The data are collected by an IBM XT computer that displays in real time on a TV monitor each scan line as it is produced. Each line contains 512 data points with the data displayed at 256 gray levels. The first scan line appears at the bottom of the monitor with each succeeding scan line appearing above the previous one. After 512 lines are displayed, the video data (256 kilobytes) are stored in a 20-megabyte hard disk drive and the screen is blanked. Each run can be programmed to last for a given number of screens. After completion, a relay system turns off the high-voltage supplies and the laser.

After a run is completed, the image data are recalled to the screen and semiautomatically converted to sequence data. This is facilitated through the use of a computer mouse. The left and right side of each lane are defined and the lane assigned a base type—A, adenine; T, thymine; G, guanine; C, cytosine—according to the loading format. Four lanes are defined for each sample with conversion accomplished one sample at a time. After lane definition, a bar cursor can be moved horizontally from one lane to another and moved vertically a programmable distance corresponding to the band-to-band distance. When the bar cursor is positioned over the band in the sequence, a key on the mouse is depressed which automatically writes the base type to the sequence data file as well as visually marking the selected band. A cassette tape drive is used to archive video data after the sequence is determined.

2.7. Data Analysis

The University of Wisconsin Genetics Computer Group (GCG) software is used for data analysis. This software program requires the use of a mainframe VAX computer and provides periodic updates of data bases such as GenBank and EMBL. A GCG file is created for each run and stored on the VAX. Since this laboratory has been involved in some service functions, the data can be directly transferred through the VAX to the files of campus users. Each user can access his/her files through his/her own terminals. For service users off-campus, the data are transferred via computer mail if they are part of BitNet or InterNet. Otherwise, the data are copied onto computer disks and the disks mailed.

The function of this laboratory is to determine sequence and not to do extensive DNA analyses and comparisons. This function is left up to the individual investigator who receives primary sequence data. However, simple programs are used to determine the accuracy of our data. Accuracy is determined by three methods. Sequence data generated from previously sequenced segments are compared to published data. For example, we can compare the data produced by our sequencing method with those already deposited in GenBank. Second, we compare multiple runs of multiple isolations and multiple runs of single isolations. This can be quickly done by using the GCG Gap program. Third, we frequently sequence clones whose inserted DNA is oriented in opposite directions. In these cases, we use the GCG Reverse program to determine the extent of homology between the paired sequence data.

3. RESULTS

Over 400,000 bases have been sequenced since mid-1986 using the method described in this chapter. Currently, the average number of bases per sample per run is greater than 500. The technique is routine and the results have high fidelity. In this section, the kinds of results that can be expected with this technique and the resultant advances in sequencing strategies will be discussed.

3.1. Raw Data

As described in Materials and Methods, the data are collected scan line by scan line, filling successive TV screens. Each run usually consists of 35 screens, although a run can be programmed to collect fewer than 35 or greater than 35 screens of data. Each run contains 4 samples (16 lanes).

These data (4 samples, 35 screens) are handled as a single file by the computer. The data can be continuously scrolled so that the junctions between screens are eliminated. A particular region can be placed at any position on the monitor.

The data look like a negative of an autoradiogram. The bands are bright spots on a black or gray background (Fig. 4). The bands tend to appear thicker than the bands on an autoradiogram. Video printouts of screens at approximately one-third the actual size will produce a montage of 500 bases/sample that is 8.5 feet long (Fig. 5). The bands do not have equal brightness, which is also similar to a standard autoradiogram.

To read the sequence, one uses the same rules applicable to autoradiograms (Hindley, 1983). For example, when there are two or more Cs in a row, the first C band is dim and the second C band is bright. This is typical when the Klenow enzyme is used.

Band dimensions change throughout a run. The widths of the bands increase as the lengths of the DNA fragments being imaged increase. This occurs because the display is in the time domain rather than the spatial domain. Actually, in space, the bands get closer and closer to each other (Brumbaugh *et al.*, 1988). The shorter DNA fragments, which form the

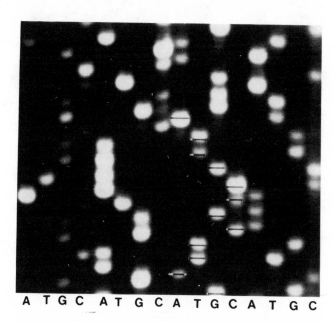

A T G C   A T G C A T G C A T G C

FIGURE 4. A typical screen of four samples (16 lanes) at 130 bases. The right center sample has cursor lines on the bands that have been written to the computer. Reprinted from *Proc. Natl. Acad. Sci. USA* **85**:5610–5614 (1988).

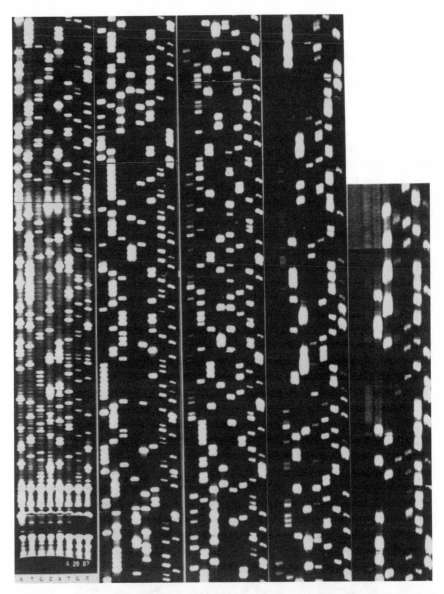

FIGURE 5. Montage of 42 overlapping videoprints of a two-sample run (eight lanes) of 500 bases. The output of the run begins at the bottom of the left panel. The left sample (left four lanes in each panel) was a segment from an alcohol dehydrogenase (ADH) allele of maize. The maize sample was AT-rich and thus the A and T bands of longer strands were weak and not detectable because the standard deoxy/dideoxy "cocktail" was used. The right sample is the pBR325 test specimen. Reprinted from *Am. Biotechnol. Lab.* **6**(6):14–22 (1988).

bands that are imaged early in a run, pass the detector fairly rapidly. Thus, the number of scan lines per band is small and the bands tend to appear thin. The longer DNA fragments, which make up the bands that occur later in a run, pass the detector much more slowly because they are moving through the gel at a slower rate. More scan lines are required to image these bands so they appear wider on the monitor.

The theoretical limiting parameter in resolving bands of longer DNA fragments is laser spot size. The laser excitation area must be able to distinguish one band from another as bands become closer and closer to each other spatially. An interesting phenomenon occurs, however, which becomes advantageous when reading the data produced by longer fragments. The band center-to-band center distances remain relatively constant throughout a run that is programmed at 20–22 W. Thus, one can discern the number of bands in regions where the band edge-to-band edge distances are quite blurred by knowing the band center-to-band center distances. This phenomenon was used to advantage in writing the image analysis software for sequence reading.

Another phenomenon that occurs as DNA fragment length increases is the degree of band distortion due to loading well malformations and/or loading artifacts. Again, this occurs because of the slower migration of the longer fragments. For example, bands migrating from a slightly convex loading well will be reasonably straight at the beginning of a run but will become increasingly more and more convex as the run continues. This creates sequence reading difficulties because band location is ambiguous. If the bottom of a well is punctured by the loading pipette, then a concave artifact occurs later in the run. A computer program has been written which addresses these problems.

Each run of raw data occupies about 10 megabytes of memory. When read, these data convert into 2000–2500 bases for the four samples. The amount of data generated per run becomes a storage and retrieval problem that will be addressed in a later section.

3.2. Chemistry

Accurate measuring and pipetting are imperative if one is to obtain consistent sequencing results. This is true not only for automated sequencing but also for standard autoradiographic sequencing. When 4–8 reactions are run simultaneously, a total of 16–32 tubes or wells contain the final reactions. It is easy to lose track of one's pipetting regimen, particularly if one is interrupted. This problem has been addressed in the literature (Pramik, 1988; Wilson *et al.*, 1988; Frank *et al.*, 1988; Landegren *et al.*, 1988).

Through the use of robotics, programmable pipetting can decrease variability and mistakes.

Accurate measurements are critical when dealing with the 1-μl mixtures of dNTPs and ddNTPs. Some variability in the amount of template DNA can be tolerated. The more accurate the measurement of template DNA, the more consistent are the results. The concentration of each sample of template DNA should be determined. The concentration of DNA at 260 nm can be measured using microcuvettes (50- to 100-μl capacity) which use less sample.

It is necessary to alter reaction mixtures to compensate for the composition of the DNA if it is AT- or GC-rich as discussed in Section 2.4. This is why an excess amount of sample DNA is prepared. Enough DNA is available for several reactions with varying concentrations of A and T or G and C pools. Figure 5 (left four lanes) illustrates how the longer fragments ending in A or T are underrepresented in an AT-rich sample when compared to an "average" sample (right four lanes). Both samples were prepared using standard reaction mixtures. As a first approximation, A and T dNTP pools are increased to 4 μl each in AT-rich samples while C and G dNTP pools are increased to 4 μl each in GC-rich samples. This usually corrects for the underrepresentation of the longer fragments.

One of the artifacts of sequencing with Klenow enzyme is the appearance of "ghost" bands. These are bands that appear in more than one lane in the same lateral position. When the ratio of template DNA to the nucleotide pools is excessive, "ghost" T bands and very bright bands of the shorter fragments appear. This makes interpretation difficult and reduces the number of sequenceable bands. The template DNA uses up the nucleotide pools and produces only the shorter fragments. Knowing the concentration of template DNA and using a consistent amount for each reaction reduces this problem as mentioned previously.

Compressions in GC-rich regions become a problem unless various deoxyguanine analogues are used. The 7-deazaguanine analogue is very effective in GC-rich regions. The band spacings become normal when this analogue is used in place of dGTP.

In some extremely GC-rich regions, one may observe multiple bands in every lane and/or "stair step" patterns. This is a consequence of secondary structure and the sequence indicated probably has skipped through the region of the compression or loop structure that formed in the DNA. Sometimes skipping through a loop structure will be indicated by a band in every lane. The sequence both before and after this position is usually correct. There are, however, bases that have not been sequenced. As mentioned in Section 2.4, the use of Taq polymerase at an elevated temperature is a way of eliminating the secondary structure and getting accurate sequence data through the region. If poly(GC) linkers have been used, one will be unable to determine precisely the number of bases in the linkers and perhaps not be able to successfully sequence from the linker region into the DNA that one wishes to sequence.

In summary, the chemical reactions will produce consistent results if one is careful with pipetting and in accurately measuring the amount of template DNA used in the reaction. This does not completely eliminate the necessity for second or third runs if the DNA composition with regard to proportions of bases or palindromic sequences is unknown prior to actually sequencing the fragment itself.

3.3. Gels

The manufacture of consistent gels is important when carrying out repetitious sequencing. Good gels are produced when the glass plates are prepared as described in Section 2.5 and when fresh clean components are used. The pouring of the gels is rather routine if one is careful to adjust the glass plates so that no leaks occur. Of particular importance is the well morphology as noted in Section 3.1. Well imperfections can cause problems in sequencing the longer DNA fragments. The treatment of the glass plates with binding silane helps bind the acrylamide to the glass plate so that when the comb is removed, the well morphology is crisp. Four percent acrylamide gels could be very useful in rapid sequencing but methods need to be developed so that well morphologies remain crisp after comb removal.

3.4. Detection Limits

The lower limit of detection of difluoresceinated DNA using this system has been determined. Dilutions of the difluoresceinated primer (54, 27, 5.4, 2.7, 0.54, 0.27, 0.054, 0.027 fmoles) were loaded and electrophoresed. The lowest detectable band was 0.054 fmole (Fig. 6). This is approximately 3 $\times 10^7$ molecules in the band. (The number of molecules actually detected is considerably less since only a fraction of the molecules in the band are illuminated.) The 0.54 and 0.27 bands most closely resemble the average band as seen on the TV monitor. The lower limit of detection could be reduced even further by using a higher-powered laser or by attaching more fluoresceins to the primer.

3.5. Image Analysis Software

Three software enhancements improve the accuracy and length of sequenceable DNA per run. The first is the programmable "jump" in the

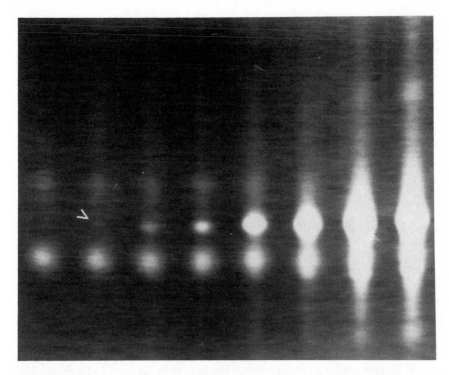

FIGURE 6. Videoprint of screen showing limits of detection. Lanes from left to right: 0.027, 0.054, 0.27, 0.54, 2.7, 5.4, 27, and 54 fmoles, respectively. The middle band is the difluoresceinated primer. The forward and trailing bands are low-molecular-weight loading buffer constituents. The 0.054 fmole band (arrowhead) contained the lowest amount of detectable fluorescence. Reprinted from *Am. Biotechnol. Lab.* **6**(6):14–22 (1988).

cursor that marks the selected bands. The "jump" gives a predicted location for the next band. This "jump" is based on the fact that band center-to-band center distances remain relatively constant throughout a run as previously described. In a region where there is a group of three or more bases of the same type, the programmable "jump" accurately determines the number of bases in the region, even if the bands are blurred together. The "jump" distance can be adjusted ("tweaked") from screen to screen to accommodate the slight variations in band center distances that occur throughout a run. The programmable "jump" is also useful toward the end of a run if one of the lanes ceases to have detectable bands. In this case, the programmable "jump" indicates that a band should appear in that lane at a particular spot. We do not use sequences derived in the latter manner as final data. The data are surprisingly accurate and informative, however, as compared to a second run when bands are present in all lanes.

A second computer enhancement enables the user to realign the lanes with regard to curvatures that may occur. As previously described, well morphology may cause concave, convex, or S-shaped bands. It is difficult to determine the precise location of these bands with regard to their neighboring bands because of these distortions. The computer program allows the operator to change the alignment of the distorted lane to straighten the bands, thus accurately determining their position with regard to bands in adjoining lanes. This is particularly useful when sequencing longer DNA fragments.

A third, lesser-used software program is background subtraction. This allows the removal of background fluorescence which can be due to the accumulation of residual fluoresceinated DNA in the lanes (i.e., a dirty track) or background fluorescence in the gel itself. Some batches of acrylamide or urea may produce more background than others. A non-band-containing region is blocked out and its brightness value removed from the entire screen. This enhances the bands that are present. It is the least used computer program because background fluorescence is usually not a problem.

Allowing human intervention to be a part of the overall data interpretation greatly enhances the accuracy at a minimal increase in human resource requirements. The individual can subconsciously take into account band trends over the course of the run. As an example, the loss of bands in a specific lane can be taken into account when sequencing areas where no band is present in any lane at an expected band location. The base type that had faded out is more than likely the correct choice at that location. Another example is when weak bands are selected over strong ghost bands in the same horizontal zone (but in different lanes) when it is known that a particular lane suffers from ghost banding. Selection of bands in this "ghosting" lane is limited to the case where no bands in the other lanes are present in that particular horizontal zone.

3.6. Error Rates

The rate of error using this sequencing system was determined in three ways. The first way was to sequence the previously described piece of pBR325 under different conditions with multiple isolations and compare it to the published sequence. We found that at position #5696, our isolate contained a thymine instead of the cytosine of the published sequence. Since this was consistent in all runs, we used the thymine as the normal sequence

TABLE I. Average Cumulative Reading Errors Using
Test Specimen (6% Acrylamide, 8 M Urea, 50°C)

	Bases					
	1–300	to 400	to 450	to 500	to 550	to 600
Average number of errors[a]	0.35	1.02	1.69	3.33	5.07	5.94
n (number of trials)	139	135	122	98	68	34
Average % errors	0.12	0.26	0.38	0.67	0.92	0.99

[a] Average number of errors per sample analyzed to base number indicated.

for our sample. Table I displays the average cumulative reading errors using this test specimen. As can be seen, the error rates are virtually nonexistent to base 400, and in the region from 500 to 600 bases, the reading errors are less than 1%.

The second method of determining error rates involved doing multiple isolations and multiple runs of the same fragment of DNA, when the DNA sequence was unknown. The number of differences between sequences of these multiple isolations and multiple runs was determined. This kind of information is displayed in Table II, which compares separate isolations of maize alcohol dehydrogenase fragments read to 500 bases. As can be seen, the number of discrepancies was 0.13%.

A third way of determining error rates was to compare the sequences from two clones whose fragments were oriented in opposite directions. One sequence should be the reverse complement of the other. The two sequences were determined and analyzed for complementation. In some cases, it has been possible to sequence from the 5' cloning site to the 3' cloning site and on into the vector. This method not only determines the error rates but helps

TABLE II. Comparison of Sequence Data Obtained from Separate Isolations
of Maize Alcohol Dehydrogenase Fragments Read to 500 Bases

Clone designation	No. of clones	Total no. of differences between clones
18N	3	0
18Q	3	3
19D	2	3
19N	3	1

```
 51 TTCATATAGGGCAGGACCACCAGAGGAGACCCTTCACTCCTCAGGCAGAT 100
                         ||||||||| ||| || |||||||||
613 .....................GAGAGACCCTTAACTTCTTAGGCAGAT 587

101 TCCCAAGAGTGAGAGAACATTGGGATGGCTGAAGTCTTTCATGATGATTC 150
    ||||||||||||||| ||| ||||||||||||||||||||||||||||||
586 TCCCAAGAGTGAGAG.ACAATGGGATGGCTGAAGTCTTTCATGATGATTC 538

151 CCTCAGTCAGAAACTGGGAGACCTCTTCTATATCTGTGATTCTATTCAAG 200
    ||||||||||||||||||||||||||||||||||||||||||||||||||
537 CCTCAGTCAGAAACTGGGAGACCTCTTCTATATCTGTGATTCTATTCAAG 488

201 GATTTCACAGCACAGTGAATTTTCTTTCCGTCATTGTCCAGCAAAGTCCC 250
    ||||||||||||||||||||||||||||||||||||||||||||||||||
487 GATTTCACAGCACAGTGAATTTTCTTTCCGTCATTGTCCAGCAAAGTCCC 438

251 ATGATAGACACAGCCAAAATGCCCTCTTCCTATGACTTCATTGAAATGCA 300
    ||||||||||||||||||||||||||||||||||||||||||||||||||
437 ATGATAGACACAGCCAAAATGCCCTCTTCCTATGACTTCATTGAAATGCA 388

301 CAATCAGGCTGCTGGGTCCAATCACAACGTGCTGAACTGCTTGGACCAGC 350
    ||||||||||||||||||||||||||||||||||||||||||||||||||
387 CAATCAGGCTGCTGGGTCCAATCACAACGTGCTGAACTGCTTGGACCAGC 338

351 TCTGGATTTAGAGCACTGAGGTCAATGTGAACAGTATTTTGTAGTAATGG 400
    ||||||||||||||||||||||||||||||||||||||||||||||||||
337 TCTGGATTTAGAGCACTGAGGTCAATGTGAACAGTATTTTGTAGTAATGG 288

401 GCTGGATATATCAGAGTCTCCACTCGTCAGGATAGGGGACAGGTCTGTCA 450
    ||||||||||||||||||||||||||||||||||||||||||||||||||
287 GCTGGATATATCAGAGTCTCCACTCGTCAGGATAGGGGACAGGTCTGTCA 238

451 GAGGATATTGCACTTGTCTGCATGCTCCATTCTGAGAGGAGTTGGGAAAC 500
    ||||||||||||||||||||||||||||||||||||||||||||||||||
237 GAGGATATTGCACTTGTCTGCATGCTCCATTCTGAGAGGAGTTGGGAAAC 188

501 TGGTCTTCTGGAAAAGTAGCTCTGTAGTCTACAGACTCATTTGAAACCAT 550
    ||||||||||||||||||||||||||||||||||||||||||||||||||
187 TGGTCTTCTGGAAAAGTAGCTCTGTAGTCTACAGACTCATTTGAAACCAT 138

551 CTCTGTAGTTGGACTTACACTTCGGGCCACTTACAAGCCTATCCAAATGA 600
    |||||||||||||||||||||||||||| |||||||||||||||||||||
137 CTCTGTAGTTGGACTTACACTTCGGG.CACTTACAAGCCTATCCAAATGA 89

601 GGAGTGTGTACTTTTGCGCCATAGCG..................... 626
    ||||||||||| ||||| |||||||
 88 GGAGTGTGTACTCTTGCGTCATAGCGAACTAATTCACTGCCCAGATCCTC 39
```

FIGURE 7. Sequencing results of a fragment cloned in opposite orientations. The reverse complement was printed for the second fragment. Of 551 bases analyzed, there are eight differences (1.45% error) and a continuum of 456 bases without any discrepancies.

resolve any differences that might occur in the two sequences, since the earlier part of one sequence is the later part of the other sequence. Figure 7 displays one such experiment which had an error rate of 1.45% in 551 bases with no discrepancies in a 456-base continuum.

Using the same strategy, one is able to take fragments of around 1 kb in length which are oriented in opposite directions and compare the sequence in the overlapping regions ($\cong$50–100 bases) and get preliminary sequence information. With good-quality runs the overlapping sequence data are comparable in the region from 500 to 600 bases. One such comparison is shown in Fig. 8, where the region from bases 500 to 588 was compared with the reverse complement of an overlapping clone in the region of bases 500–633. Only ten discrepancies in the 90-base region of overlap occurred.

The errors are of three types. Some errors are due to conditions involving the enzymatic and chemical reactions such as "ghost" bands which have already been described. Such errors are not due to the detection system and would be present if standard autoradiographic methods were used. The second type of error occurs near the end of a sequence when band separations become less distinct and difficult to interpret. In these cases, the programmable "jump" in the computer program allows a rather accurate determination of the number of bases present. The third type of error is human error—misread band patterns. Errors due to chemical conditions usually cause substitutions. The errors which occur at the end of a run where the number of bases is questionable cause insertions or deletions. Human errors are probably equally divided between substitutions and deletions or insertions.

3.7. Data Storage

In the standard sequencing laboratory, autoradiograms are filed so that they can be reviewed. Stored data from each run of an automated sequencer

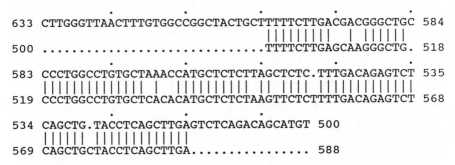

FIGURE 8. The overlap produced by sequencing a piece of DNA approximately 1 kb long from either end. The bases were determined by sequencing two clones each with the piece oriented in opposite directions. From base 500 and greater in each clone an overlap of 90 bases was determined containing only ten discrepancies.

should also be available for review. A computer tape drive is used to archive the raw data from each run. Thus, instead of having a file containing autoradiograms, we have a file containing computer tapes. Although not used extensively, the archiving system has been useful on a number of occasions when a particular region of a particular run has been retrieved, examined, and an ambiguity resolved. It is essential that such raw data be archived so that in subsequent years, a particular run can be reviewed when a sequence discrepancy is reported in the data banks and/or literature.

4. DISCUSSION

4.1. Comparisons of Automated Sequencing Methods

Gel readers are being developed (Elder *et al.,* 1986; West, 1988) which determine sequence from standard autoradiograms. Gel readers do not collect or interpret data in real time but determine sequence by the batch method in the spatial domain. Some of the algorithms used to determine sequence from autoradiograms should be applicable to reading raw data from automated sequencers.

Automated sequencers which collect data in real time can be divided into two broad categories. One category still uses radioactivity as the signal. Instruments have been developed by Hitachi (Nagai *et al.,* 1987), EG & G (Page, 1988), and Genofit (Smith, 1988). Single dimension curves are produced indicating band positions using scintillation detectors for each of the four lanes per sample.

The second category of real time automated DNA sequencers uses variations of fluorescently marked DNA. There are four different kinds of fluorescent real time automated DNA sequencers. In addition to the method described in this chapter, two other methods use fluorescently tagged primers. The instruments developed by Ansorge *et al.* (1986, 1987) and Kambara *et al.* (1988) use a primer whose fluorescence is excited by passing a laser beam between the two glass plates that mold the gel. The instrument developed by Smith *et al.* (1986) uses four different fluorescent dyes to mark the DNA. Each of the four dyes marks a specific base type. Therefore, a single lane can determine the information for the sequence of a sample as opposed to the four-lane format used by the other methods. Prober *et al.* (1987) have developed a machine that depends on fluorescently tagged dideoxy analogues. Four different dyes are used so a single lane can determine the sequence of a sample.

These four methods can be compared with regard to several parameters (see Table III). All the methods except the one described in this chapter have only single fluorescent reporter groups per DNA molecule. Because the

TABLE III. Comparison of Fluorescently Based
Automated DNA Sequencing Methods

	Smith *et al.* (1986)	Ansorge *et al.* (1986, 1987)	Prober *et al.* (1987)	Kambara *et al.* (1988)	This chapter
No. of different fluorescent emitters	4	1	4	1	1
No. of fluors/ DNA molecule	1	1	1	1	2
No. of primer hybridizations	4	1	1	1	1
No. of elongation reactions	4	4	1	4	4
Excitation method	Scanning	Side entry	Scanning	Side entry	Scanning
Background reduction by chopping	No	No	No	No	Yes
Type of display	One-dimen-sional curves	One-dimen-sional curves	One-dimen-sional curves	One-dimen-sional curves	Two-dimen-sional TV display
Reading	Auto-mated	Auto-mated	Auto-mated	Auto-mated	Semi-auto-mated

primer used in the method described in this chapter contains two fluorescein molecules, greater signal per DNA molecule is produced.

The method described in this chapter uses only one type of fluorescent dye, as do the methods developed by Ansorge *et al.* (1986, 1987) and Kambara *et al.* (1988). Both Smith *et al.* (1986) and Prober *et al.* (1987) use four different dyes, one to indicate each of the four different base types. This is both an advantage and a disadvantage. The advantage is that a single lane can accommodate the data for one sample. The disadvantage is that the computer analysis of the data requires the separation of the wavelength and molecular weights of the different reporter molecules used, at least for the Smith *et al.* (1986) method.

The method described in this chapter and the method developed by Ansorge *et al.* (1986, 1987) and Kambara *et al.* (1988) require one hybridization. The method developed by Smith *et al.* (1986) requires four hybridizations, one for each base type. One of the distinct advantages of the Prober *et al.* (1987) methodology is that all of the reactions can take place in a single tube because the dideoxy analog dyes themselves determine which base type they are indicating. The number of reactions is four for each of the methods except the Prober *et al.* (1987) method, which requires just one.

From the literature, it has been determined that chopping the laser signal to reduce or eliminate background fluorescence is unique to the system described in this chapter. In addition, the filtering out of the Raman band of water, which has been done by both this method and that reported by Ansorge *et al.* (1986, 1987), is essential to reduce noise.

Unique to the method described in this chapter is the two-dimensional collection and display of the data. The displays for the other instruments are simply one-dimensional curves of each of the base types rather than a two-dimensional display which resembles an autoradiogram.

4.2. Advantages and Disadvantages of This Method

The advantages of this method include the elimination of radioactivity and the presence of two reporter groups per DNA molecule which increases the signal. Only this method chops the laser beam to decrease the inherent fluorescence in the glass and the gel. As opposed to the Smith *et al.* (1986) method, only a single hybridization of primer to template is needed.

Perhaps the strongest advantage of this system is the collection of two-dimensional data. The full two-dimensional pattern of the bands and background is displayed. This allows for human interpretation to utilize all aspects of the brain's image processing capabilities, particularly the ability to incorporate the relationship of bands to each other. In addition, it allows for a better computerized, nonhuman attempt at decoding the image information through the use of two-dimensional filtering, neural network analysis, and other computerized image analysis. Information that is necessary for proper interpretation is not lost as it is in the case of comparing four one-dimensional graphs or curves.

The disadvantages of this method have not been prohibitive in the development of automated sequencing. The use of four lanes instead of one has not been a disadvantage because of the close temperature control of the lanes which eliminates smiling and other artifacts. The use of four lanes has also been augmented by the development of the software for band straightening. With the Prober *et al.* (1987) chemistry, all of the reactions can be carried out in one tube. The use of four reaction tubes is not necessarily a disadvantage because dNTP pool sizes can be varied in each reaction to accommodate the variable $A+T/G+C$ ratios of the samples being sequenced. Thus, this sup-

CONVENTIONAL

Assumptions - 300 bases / run ; 50 base overlap

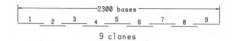

AUTOMATED

Assumptions - 600 bases / run ; 50 base overlap

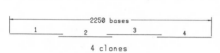

FIGURE 9. Comparisons of the number of subclones needed to span a region of DNA for sequencing using conventional means (autoradiography) and automated sequencing. The number of subclones is reduced from nine to four for a region of 2250–2300 bases.

posed disadvantage can actually be an advantage in that it permits more flexible reaction chemistries. All of the fluorescent techniques except that developed by Prober *et al.* (1987) would require the custom synthesis of different fluorescently marked primers for use in nonstandard vectors or for "chromosome walking" techniques. Any hybridizable oligonucleotide can be used with the Prober *et al.* (1987) methodology because the fluorescence is provided by dideoxy analogues.

4.3. Sequencing Strategies

Automated sequencing, like that described in this chapter, which allows longer fragments of DNA to be sequenced in a single run, significantly changes subcloning strategies. If one assumes that standard autoradiographic procedures can produce about 300 bases per sample per load per run, then to sequence a 2300-base segment of DNA with an overlap of 50 bases between samples to establish linkage would require nine clones of DNA. Only four clones would be required if 600 bases per sample were determined (see Fig. 9). At this point, we have not maximized every parameter of our sequencing reactions or running conditions. It may be possible that 1000 bases per run per sample per reaction can be achieved. This would greatly simplify the design and execution of sequencing strategies.

ACKNOWLEDGMENTS. We thank Nancy Bayne for her careful preparation of the manuscript, K. W. Lee for photographic assistance, and John Osterman for the ADH clones. The research was supported by funds from Li-Cor, Inc., and reagents from Molecular Biosystems, Inc.

5. REFERENCES

Ansorge, W., and Barker, R., 1984, *J. Biochem. Biophys. Methods* **9**:33–47.

Ansorge, W., and Labeit, S., 1984, *J. Biochem. Biophys. Methods* **10**:237–243.

Ansorge, W., Sproat, B. S., Stegemann, J., and Schwager, C., 1986, *J. Biochem. Biophys. Methods* **13**:315–323.

Ansorge, W., Sproat, B., Stegemann, J., Schwager, C., and Zenke, M., 1987, *Nucleic Acids Res.* **15**(11):4593–4602.

Ausubel, F. M., Brent, R., Kingston, R. E., Moore, D. D., Seidman, J. G., Smith, J. A., and Struhl, K., 1987, *Current Protocols in Molecular Biology,* Greene Publishing Associates and Wiley–Interscience, New York.

Barr, P. J., Thayer, R. M., Laybourn, P., Najarian, R. C., Seela, F., and Tolan, D. R., 1986, *BioTechniques* **4**(5):428–431.

Beck, S., 1987, *Anal. Biochem.* **164**(2):514–520.

Beck, S., and Pohl, F. M., 1984, *EMBO J.* **3**:2905–2909.

Biggen, M. D., Gibson, T. J., and Hong, G. F., 1983, *Proc. Natl. Acad. Sci. USA* **80**:3963–3965.

BRL, 1980, M13 cloning/dideoxy sequencing manual, Bethesda Research Laboratories, Inc.

Brumbaugh, J. A., Middendorf, L. R., Grone, D. L., and Ruth, J. L., 1988, *Proc. Natl. Acad. Sci. USA* **85**:5610–5614.

Church, G. M., and Gilbert, W., 1984, *Proc. Natl. Acad. Sci. USA* **81**:1991–1995.

Church, G. M., and Kieffer-Higgins, S., 1988, *Science* **240**:185–188.

Elder, J. K., Green, D. K., and Southern, E. M., 1986, *Nucleic Acids Res.* **14**:417–424.

Frank, R., Bosserhoff, A., Boulin, C., Epstein, A., Gausepohl, H., and Ashman, K., 1988, *BioTechnology* **6**:1211–1213.

Garoff, H., and Ansorge, W., 1981, *Anal. Biochem.* **115**:450–457.

Hindley, J., 1983, *DNA Sequencing,* Elsevier, Amsterdam, pp. 200–206.

Jablonski, E., Moomaw, E. W., Tullis, R. H., and Ruth, J. L., 1986, *Nucleic Acids Res.* **14**(15):6115–6128.

Kambara, H., Nishikawa, T., Katayama, Y., and Yamaguchi, T., 1988, *BioTechnology* **6**:816–821.

Landegren, U., Kaiser, R., Caskey, C. T., and Hood, L., 1988, *Science* **242**:229–237.

Maxam, A. M., and Gilbert, W., 1977, *Proc. Natl. Acad. Sci. USA* **74**:560–564.

Maxam, A. M., and Gilbert, W., 1980, *Methods Enzymol.* **65**:499–560.

Messing, J., 1983, *Methods Enzymol.* **101**:20–78.

Middendorf, L. R., Brumbaugh, J. A., Grone, D. L., Morgan, C. A., and Ruth, J. L., 1988, *Am. Biotechnol. Lab.* **6**(6):14–22.

Mizusawa, S., Nishimura, S., and Seela, F., 1986, *Nucleic Acids Res.* **14**(3):1319–1324.

Nagai, K., Shimida, T., Tokita, J., Watanabe, K., Nakano, R., Kanbara, H., Sumitani, T., Teranishi, Y., and Hishinuma, F., 1987, Abstracts of the 1987 Pittsburgh Conference and Exposition on Analytical Chemistry and Applied Spectroscopy, Abstract 545 (March 9–13, 1987, Atlantic City, NJ).

National Research Council, 1988, Mapping and Sequencing the Human Genome, National Academy Press, Washington, D.C.

Olsson, A., Moks, T., Uhlen, M., and Gaal, A. B., 1984, *J. Biochem. Biophys. Methods* **10**:83–91.

Page, G., 1988, *Nature* **333**:477–478.

Pramik, M. J., 1988, *Genetic Engineering News* Jan. 1988, p. 6.

Prentki, P., Karch, F., Iida, S., and Meyer, J., 1981, *Gene* **14**:289–299.

Prober, J. M., Trainor, G. L., Dam, R. J., Hobbs, F. W., Robertson, C. W., Zagursky, R. J., Cocuzza, A. J., Jensen, M. A., and Baumeister, K., 1987, *Science* **238**:336–341.

Ruth, J. L., 1984, *DNA* **3**:123 (abstr.).

Ruth, J. L., Morgan, C., and Pasko, A., 1985, *DNA* **4**(1):93.

Sanger, F., Niklen, S., and Coulson, A. R., 1977, *Proc. Natl. Acad. Sci. USA* **74**:5463–5467.

Smith, L. M., 1988, *The Scientist* July 25, 1988, p. 23.

Smith, L. M., Sanders, J. Z., Kaiser, R. J., Hughes, P., Dodd, C., Connell, C. R., Heiner, C., Kent, B. H., and Hood, L., 1986, *Nature* **321**:674–679.

Vieira, J., and Messing, J., 1987, *Methods Enzymol.* **153**:3–11.

Wada, A., 1984, *Nature* **307**:193.

Wada, A., 1987, *Nature* **325**:771–772.

West, J., 1988, *Nucleic Acids Res.* **16**:1847–1856.

Wilson, R. K., Yuen, A. S., Clark, S. M., Spence, C., Arakelian, P., and Hood, L. E., 1988, *BioTechniques* **6**:776–787.

Chapter 4

Time-Resolved Fluorescence in Biology and Biochemistry

David M. Jameson and Theodore L. Hazlett

1. INTRODUCTION

A number of articles and books have appeared in recent years describing in varying detail the theory and application of time-resolved fluorescence spectroscopy in the chemical and biological sciences (Badea and Brand, 1979; Demas, 1983; Lakowicz, 1983; Cundall and Dale, 1983; Gratton *et al.*, 1984a; Jameson *et al.*, 1984; Taylor *et al.*, 1987; Jameson and Reinhart, 1989). Our goal in this chapter is thus not to treat the subject exhaustively but rather to reiterate the fundamental considerations and to provide the reader with a feeling for the practical aspects that must be apprehended before time-resolved methods can be profitably applied to complex biological systems.

The appeal of fluorescence spectroscopy in biological research lies not only in the inherent sensitivity of the method but also in the fact that the fluorescent probe provides us with a molecular stopwatch with which we can monitor microscopic and submicroscopic events. The clock essentially starts with the absorption of a light quantum and stops with the deactivation of the excited state be that through the emission of a photon or some nonradiative process such as quenching. Only those processes manifested during the ex-

DAVID M. JAMESON and THEODORE L. HAZLETT • Department of Biochemistry and Biophysics, John A. Burns School of Medicine, University of Hawaii at Manoa, Honolulu, Hawaii 96822.

cited state lifetime will be observable and hence an accurate determination of the fluorescence lifetime is often essential for correct interpretation.

The absorption of the light quantum occurs rapidly, according to the Franck–Condon principle, in the range of femtoseconds. Thermalization processes then follow in picoseconds which leaves the molecule in the lowest vibrational level of the first electronic excited state. The deactivation of this state (A*) can be considered in terms of the corresponding rate constants:

$$-d(A^*)/dt = (k_f + k_{nr})(A^*) \tag{1}$$

where k_f and k_{nr} are associated with radiative and nonradiative processes, respectively. Integration of this expression leads to Eq. (2) which indicates that the deactivation process should correspond to a single exponential decay:

$$I(t) = \exp(-t/\tau) \tag{2}$$

where τ is the fluorescence lifetime.

In practice, however, single exponential decays are by far the exception. Especially in the more complex systems considered in biological studies the emission is likely to be more complicated. Hence, a fundamental consideration in the analysis of such systems is the nature and extent of the heterogeneity. At its simplest level, for example, we may wish to know if the observed heterogeneity results merely from a mélange of fluorophores that can ultimately be resolved into the constituents by appropriate physical or chemical interventions. Or, the heterogeneity may be a fundamental aspect of the system, arising as a consequence of environmental complexity or competing deactivating rate processes. One of the more obvious examples of this inherent complexity is the fact that almost all of the single-tryptophan-containing proteins studied to date give rise to an emission which cannot be well fit to a single exponential decay (Beechem and Brand, 1985). Various approaches for analyzing heterogeneous emissions will be considered in a subsequent section.

2. INSTRUMENTATION FOR TIME-RESOLVED MEASUREMENTS

Time-resolved fluorescence measurements are generally accomplished by either the impulse response method or the harmonic response method. In the impulse response approach the illumination function is a light pulse of short duration, ideally a delta function. Such pulses have traditionally been generated by flash lamps or pulsed laser sources (Demas, 1983) although synchrotron radiation sources have been used occasionally (Laws and Jameson, 1989). The direct decay curve, i.e., the record of fluorescence intensity versus time, is obtained using a detector system with temporal resolution such as streak cameras (Nordlund, 1988), time-sampling circuitry or, most commonly, the time-correlated single photon counting method (Badea and Brand, 1979; Demas, 1983).

In the traditional harmonic response approach, the intensity of the exciting light source is modulated sinusoidally at high frequencies, typically in the megahertz range. The fluorescence of a sample subjected to this modulated excitation function will also be modulated sinusoidally but delayed in phase, relative to the excitation, due to the finite persistence of the excited state (Fig. 1). More recently, the impulse and harmonic response methods have been used coincidentally; in this approach the excitation function is intrinsically pulsed yet the detection system takes advantage of the harmonic content of the repetitive pulsed source to utilize normal phase/modulation methods (Gratton *et al.*, 1984b; Alcala *et al.*, 1985; Gratton and Barbieri, 1986). Since our own experience has been mostly with the harmonic response (phase) method we shall focus the rest of our discourse along those lines. Detailed descriptions of the impulse response method are plentiful (see, e.g., Demas, 1983; Badea and Brand, 1979). We should note that the data treatment and analysis for frequency domain results are identical whether traditional, i.e., sinusoidal excitation modulation, is utilized or whether the pulsed excitation/harmonic content approach is used.

The first phase fluorometer appears to be that of Gaviola, described more than 60 years ago (Gaviola, 1926). At that time, Gaviola relied upon visual detection of the fluorescence and was able to determine the lifetime of

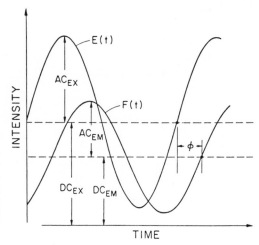

FIGURE 1. Diagram of the excitation $E(t)$ and emission $F(t)$ waveforms. The phase delay, ϕ, between the excitation and emission waveforms and the modulation ratio, $(AC_{EM}/DC_{EM})/AC_{EX}/DC_{EX})$, are used to determine the lifetime of the sample fluorescence.

rhodamine B, with good precision, using only optical delay lines and visual compensation methods. Only some years later did Duschinsky derive the fundamental theory of the phase fluorometer (Duschinsky, 1933). He demonstrated that upon excitation by light with an intensity modulated sinusoidally at angular frequency, ω, a fluorophore characterized by a single exponential decay time, τ, will emit light sinusoidally modulated at the same frequency but delayed in phase by an angle, ϕ, and demodulated with respect to the excitation (Fig. 1). The demodulation, M, is the ratio of the signal at frequency ω to the average (DC) signal, i.e., the AC/DC ratio. The relationships between the phase shift and modulation ratio and lifetime are:

$$\tan\phi = \omega\tau_p \tag{3}$$

$$M = [1 + (\omega\tau_m)^2]^{-1/2} \tag{4}$$

Hence, two independent determinations of the fluorescence lifetime, i.e., τ_p and τ_m, are available. If the lifetime does, in fact, correspond to a single exponential decay, then the phase and modulation lifetimes will be identical. If, however, the emission is due to two or more exponential decay processes, then the apparent phase lifetime will be less than the apparent modulation lifetime (Spencer and Weber, 1969). Moreover, in the case of such lifetime heterogeneity the apparent phase and modulation lifetimes will be dependent upon the modulation frequency, namely decreasing as the modulation frequency increases. In the earlier days of phase and modulation fluorometry, when only one or two modulation frequencies were readily available, this difference between the phase and modulation lifetimes was important for establishing the heterogeneous nature of the lifetime and giving an estimate of the extent of heterogeneity. With the present generation of wide-range multifrequency instrumentation (to be described in more detail later) the tendency is not to focus on the difference between the phase and modulation lifetimes as a test of heterogeneity but rather to view the variation in phase or modulation lifetimes over a wide frequency range.

The development of phase fluorometry instrumentation has been reviewed in varying detail several times (Birks and Munro, 1967; Teale, 1983; Jameson *et al.*, 1984) and we shall not reiterate it here. Rather, we shall begin our discussion with a description of the more recent multifrequency phase fluorometers, focusing on the instrumentation developed at the University of Illinois at Urbana–Champaign. The instrument described by Gratton (Gratton and Limkeman, 1983) was the first true wide-range multifrequency phase fluorometer which utilized a Pockels cell for light modulation and the cross-correlation technique, popularized by Spencer and Weber (1969) (also see Jameson *et al.*, 1984, for a discussion of cross-correlation methods and

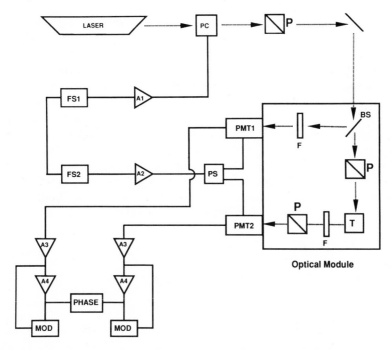

FIGURE 2. Block diagram of a cross-correlation phase and modulation fluorometer. PC, Pockels cell; P, polarizer; BS, beam splitter; F, filters; T, sample turret; PMT, photomultiplier tube; PS, power splitter; A, amplifier; FS, frequency synthesizer; PHASE, phase meter; MOD, modulation meter.

the principles behind the light modulation strategy). Figure 2 shows a block diagram of such an instrument. We should also note that, typically, "magic angle" polarizer positions are utilized: specifically, in our own instrument we use vertically polarized exciting light and observe the emission through a polarizer oriented at 55° to the vertical laboratory axis. The effect of polarized emission on lifetime measurements has been treated elsewhere (Spencer and Weber, 1970). In a typical measurement the phase delay and modulation ratio for scattered light (from glycogen or a suspension of latex particles) and the fluorescence are obtained relative to a reference photomultiplier or internal reference signal. The absolute phase delay of the fluorescence, ϕ, is then given as:

$$\phi = (\phi_R - \phi_F) - (\phi_R - \phi_S) \tag{5}$$

where ϕ_R represents the phase of the internal electronic reference signal (or the signal from the reference photomultiplier) and ϕ_F and ϕ_S represent the measured phase readings for the fluorescent and scattered signals, respectively. The phase reading is measured in degrees, typically with a precision on the order of 0.1–0.2°, over the entire frequency range which is selected by wide-range frequency synthesizers. The frequency can be changed in seconds either manually or under computer control—no additional adjustments to the light modulation hardware are required (as compared to the older Debye–Sears acousto-optic light modulators which often required extensive realignment upon each frequency change). The phase and modulation data at a number of frequencies are thus obtained and typically plotted as shown in Fig. 3. The frequency range utilized will depend upon the lifetime(s) being measured. Specifically, one usually tries to choose a frequency range such that the observed phase angles vary from about 20 to 70°. The frequency range encompassed, though, and the number of frequencies utilized will depend greatly upon the complexity of the emission process and the precision sought in the resolved lifetimes (for a discussion of these considerations see Gratton *et al.,* 1984a). In the case of single exponential decays one can, of course, have an accurate lifetime value with a single frequency measurement. Since phase measurements are inherently real-time (a single frequency measurement can be obtained in a matter of seconds), the method has great value in those cases where experimental circumstances require a rapid assessment.

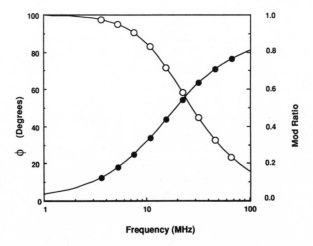

FIGURE 3. Data points, phase (●) and modulation (O), and curves expected for a species with a single, 10-nsec lifetime.

3. DATA ANALYSIS

3.1. Discrete Exponentials

The analysis of heterogeneous emissions in terms of the component lifetimes and fractional contributions has attracted a tremendous effort in the last two decades. In the early days of phase fluorometry the inability of a single frequency phase measurement to discern the complexity of the emission contributed to the popularity of the emerging pulse methods. The possibility of using multifrequency phase and modulation measurements to analyze for lifetime heterogeneity was recognized (Spencer and Weber, 1969) but few examples of such approaches appeared; some of this early work has been reviewed (Jameson et al., 1984). When phase and modulation values differ (keeping always in mind that the precision of the measurement will dictate whether this difference has significance), then the data must be analyzed according to a scheme more complex than single exponential. Analysis of a double exponential emission yields three independent parameters, i.e., two lifetime values and one fractional intensity contribution. Emitting systems that may be appropriately described by a double exponential scheme include the case of two independently emitting species, systems that exhibit excited state reactions between well-defined states and systems exhibiting energy transfer between two species. It is important to realize that, except for the case of two actual independent emitting species, the parameters which are associated with the two-component decay analysis cannot be assigned to particular molecular entities.

An elegant analytical solution for determining N lifetime components and their respective fractional contributions, given N modulation frequencies, was derived by Weber (Weber, 1981). This algorithm was, in fact, the method of choice when only two or three frequencies were available. As wide-range multifrequency instruments were developed, it was found that Weber's algorithm required very stringent precision as the number of frequencies increased (Jameson and Gratton, 1983). This aspect of the analytical solution led to the general adoption of nonlinear least-squares fitting routines (Jameson and Gratton, 1983; Jameson et al., 1984); nonetheless, Weber's algorithm was an important step since it provided a mathematically rigorous approach and gave an additional impetus to the field.

In the nonlinear least-squares approach, the goodness of fit of the measured phase and modulation data to a particular model (e.g., single or double exponential) is judged by the value of the reduced chi-square (χ^2) as defined by:

$$\chi^2 = \sum \{ [P_c - P_m/\sigma^P] + [(M_c - M_m)/\sigma^M] \}/(2n - f - 1) \qquad (6)$$

where the sum is carried over the measured values at n modulation frequencies and f is the number of free parameters. The symbols P and M correspond to the phase shift and relative demodulation values, respectively, while the indices c and m indicate the calculated and measured values, respectively. σ^P and σ^M correspond to the standard deviations of each phase and modulation measurement, respectively. In present-day phase fluorometers, the standard deviations of the measured phase and modulation values are essentially independent of the frequency and hence it is convenient to fix a common value for these constants; typical values chosen are in the range of $0.2°$ for the phase angle and 0.04 for the demodulation ratio. The choice of these values and the actual standard deviations in a particular data set will thus clearly affect the calculated χ^2 value. Hence, the χ^2 for a given fit includes an arbitrary factor and the important parameter is the change in the observed χ^2 value upon fitting different decay models rather than the absolute χ^2 value. The calculated values of phase and modulation are obtained using the equations:

$$P = \tan^{-1}[S(\omega)/G(\omega)] \tag{7}$$

$$M^2 = S(\omega)^2 + G(\omega)^2 \tag{8}$$

where the functions $S(\omega)$ and $G(\omega)$ have different expressions depending on the fitting model utilized.

For the fit using a sum of exponentials, the functions $S(\omega)$ and $G(\omega)$ are given by

$$S(\omega) = \sum f_i\omega\tau_i/(1 + \omega^2\tau_i^2) \tag{9}$$

$$G(\omega) = \sum f_i/(1 + \omega^2\tau_i^2) \tag{10}$$

$$\sum f_i = 1 \tag{11}$$

where the index i depends on the number of exponentials used for the fit; f_i is the contribution to the steady-state fluorescence of the ith component; τ_i is its lifetime and ω is the angular frequency of light modulation.

An example of multifrequency phase and modulation data corresponding to a two-component system is given in Table I (Valat *et al.,* 1988). One notes that the phase lifetimes at any given light modulation frequency are always less than the modulation lifetimes and that both phase and modulation lifetimes decrease as the light modulation frequency increases. This particular data set corresponds to a mixture of protoporphyrin IX and photoprotoporphyrin in dioxane; a discrete three-component analysis yields lifetimes of 14.4 and 6.1 nsec when the third component is fixed at 0.001 nsec. The fractional intensities associated with these components are 0.535, 0.459, and 0.006, respectively, and the χ^2 is 0.816. One often finds that introduc-

TABLE I. Lifetime Parameters: Protoporphyrin IX
and Photoprotoporphyrin in Dioxane

Frequency (MHz)	Phase (degrees)	Modulation	τ_{Phase} (nsec)	τ_{Mod} (nsec)
10	31.14	0.821	9.62	11.08
15	39.94	0.708	8.88	10.58
20	46.57	0.613	8.40	10.25
25	52.11	0.540	8.18	9.92
30	56.05	0.476	7.88	9.79
35	59.81	0.433	7.82	9.47
40	62.45	0.392	7.63	9.35
45	64.98	0.356	7.58	9.29

tion of a small fraction ($<1\%$) of a very short component, attributable to scattered light or a Raman contribution, leads to significantly improved fits. We should also comment that the fractional intensities do not correspond to mole fractions; one must know the relationship between quantum yield and lifetime in the particular system at hand before this extrapolation can be made.

3.2. Distributions

The discrete component analysis of fluorescence lifetimes assumes that all the radiating fluorophores decay with a well-defined set of a few lifetimes. A more recent approach to analysis of fluorescence decay data is the continuous distribution method (Alcala *et al.*, 1987). In the distributional approach, one assumes that environmental complexities give rise to a spread of lifetime values that can map to a specific distribution, e.g., Lorentzian or Gaussian distributions, of fluorophore environments. The theory and rationale underlying this approach has been presented in detail (Alcala *et al.*, 1987) and here we shall give only a brief overview.

For the fit using a continuous distribution of lifetime values, the $S(\omega)$ and $G(\omega)$ functions are given by

$$S(\omega) = \int f(\tau)\omega\tau/(1 + \omega^2\tau_i^2)d\tau \qquad (12)$$

$$G(\omega) = \int f(\tau)/(1 + \omega^2\tau_i^2)d\tau \qquad (13)$$

$$\int f(\tau) = 1 \tag{14}$$

where $f(\tau)$ is an arbitrary function. In the case of a Lorentzian distribution, the function, $f(\tau)$, utilized is:

$$f(\tau) = A/[1 + 4(\tau - \tau_0)^2/W^2] \tag{15}$$

where A is determined by the normalization condition, τ_0 is the center of the Lorentzian, and W is its full width at half-maximum.

One of the important motivations for the distributional approach to lifetime analysis arose from the difficulty in rationalizing the observed decay schemes for single-tryptophan proteins. Specifically, it had been noted that only rarely could the lifetime of a single-tryptophan-containing protein be satisfactorily fit to a single-exponential decay scheme. In some cases, three- and four-component decay schemes were required to obtain a fit judged adequate by traditional statistical criteria. On the other hand, some of these cases could be well fit to a simple distributional model. An example of such a fit in a fluorophore/protein system is shown in Fig. 4 where methylanthraniloyl-GDP is bound to the N-ras p21 protein product (material courtesy of J. Eccleston, S. Neal, and M. Webb, MRC Laboratories, Mill Hill, London) (Neal *et al.,* 1988). In this case, comparable fits were found using either a two-component discrete analysis or a one-component Lorentzian distribution analysis. The distributional analysis has more recently been applied to the area of biomembranes where the decay scheme of fluorescent membrane probes, most notably DPH, has been analyzed in terms of simple distributions (Fiorini *et al.,* 1988; Williams and Stubbs, 1988). The areas of applicability and the interpretative value of the distributional analysis approach are

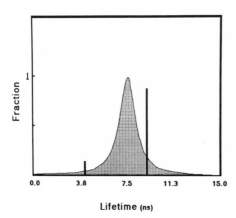

FIGURE 4. Continuous distribution (Lorentzian) lifetime analysis for methylanthraniloyl-GDP bound to the N-ras, P21 protein product (20°C). Calculated lifetime and full width at half-maximum were 7.7 and 1.1 nsec, respectively. The vertical dotted lines represent a two-component discrete lifetime fit to the same data: $\tau_1 = 8.5$ nsec, 86% of the intensity, and $\tau_2 = 3.9$ nsec. The χ^2 values were similar in both cases.

presently being examined and extensive research efforts along these lines are ongoing.

3.3. Global Analysis

In the last few years a new approach for analyzing fluorescence data—called *global analysis*—has been introduced (Beechem *et al.*, 1983; Knutson *et al.*, 1983; Beechem and Gratton, 1988). Conceptually, global analysis involves the simultaneous fitting of individual data sets having parameters in common. This treatment exploits the information on common parameters in a least-squares fit to a multidimensional data surface while the unlinked variables are fit locally to their specific data subset. A general algorithm has been developed as a framework for including mathematical expressions of physical models to be fit by the nonlinear least-squares routines. Functions introduced to date include those for fluorescence decay (discrete and distribution analyses), anisotropy decay (discrete and distribution analyses), Stern–Volmer quenching, energy transfer, and wavelength dependence, among others. The power of this approach is enhanced by its ability to link parameters among a wide range of related experiments, and the method's flexibility in adding new expressions to the core program.

An example of a "global" resolution of heterogeneous decay from a multitryptophan protein system is provided by the work of Beechem *et al.* (1983) on horse liver alcohol dehydrogenase. In this system, the intrinsic protein fluorescence arises from two tryptophan populations which have different lifetimes and spectra. By linking lifetime data (phase and modulation) obtained at different emission wavelengths, Beechem *et al.* were able to resolve the two emitting components with a higher precision than possible with conventional approaches. Similar gains have been made using global analysis routines in quenching studies (Eftink and Wasylewski, 1989) and anisotropy decays (Beechem *et al.*, 1986) as well. Development of the global analysis method is an important addition to our arsenal of analytical tools, greatly enhancing the information available from fluorescence methodologies. Global analysis software is presently available from the Laboratory for Fluorescence Dynamics at the University of Illinois at Urbana–Champaign.

4. ANISOTROPY METHODS

4.1. General Comments

Anisotropy methods are used to examine the rotational dynamics of biomolecules by exciting fluorophores with plane polarized light and then measuring the extent of depolarization of the emission, with either static (steady-state) or dynamic (time-resolved) methods. Information on dynamic aspects of macromolecules from anisotropy measurements has been extremely useful to cell biologists and biochemists in detailing information on membrane fluidity, protein–ligand interactions, and general information on macromolecular shape and structure. Other physical methods such as light scattering, ESR, NMR, gel filtration, and ultracentrifugation also provide information on some aspects of shape and motion but often require relatively high concentrations of material, or must be performed under thermodynamically nonequilibrium conditions. All of these techniques have their specific advantages and disadvantages but few have the sensitivity provided by fluorescence anisotropy. However, application of anisotropy methods, while appearing simple and straightforward, can be fraught with experimental artifacts; the literature is replete with examples of both correctly and incorrectly designed studies. Our goal here is to provide the reader with a basic background and the necessary references to evaluate intelligently whether anisotropy might be useful for the research problem at hand and to avoid some of the more common errors. We will discuss anisotropy beginning with general theory and basic instrumentation followed by various practical considerations in experiment design.

4.2. Steady-State

Similar to other fluorescence approaches, anisotropy is measured at right angles to the exciting light, but unlike other methods the exciting light is usually polarized and the emission is measured for the extent of polarization. If we define the laboratory plane as the x, y plane with the z axis normal to this plane, then the excitation light is to be polarized along the z axis (Fig. 5). The emission is then measured for light polarized along the z axis, parallel to the orientation of the exciting light, and for light perpendicular to the z axis. The total fluorescence is the sum of the polarized components along all three axes,

$$F = I_x + I_y + I_z \qquad (16)$$

where F is the total sample fluorescence and I_x, I_y, and I_z are the fluorescence along the three axes. In the case of polarized excitation along the z axis, $I_x = I_y$ and

$$F = I_z + 2I_x \qquad (17)$$

Commonly, the directionalities of I_x and I_z are referred relative to the laboratory axes. The nomenclature can be confusing for those not familiar with the subject; two common sets of equivalent terms in the literature include:

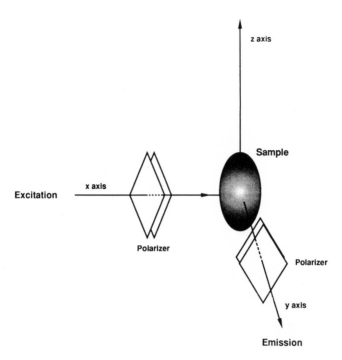

FIGURE 5. Spatial arrangement of the emission and excitation vectors with the x, y, and z axes.

$$I_x = I_{90} = I_H = I_\perp \qquad (18)$$

and

$$I_z = I_0 = I_V = I_\parallel \qquad (19)$$

where the subscripts 0 and 90 stand for $0°$ and $90°$ angles off the z axis; V and H represent vertical and horizontal; and $\parallel$ and $\perp$ are the parallel and perpendicular components, with respect to the z axis. For clarity we will use the parallel and perpendicular terminology for the remainder of this chapter.

During excitation with polarized light, the anisotropy, r, is defined as the difference between the parallel and the perpendicular emission components relative to the total sample fluorescence. Thus, from experimental values of intensity along the x and z axes, one can calculate the anisotropy,

$$r = (I_\parallel - I_\perp)/(I_\parallel + 2I_\perp) \qquad (20)$$

The term originally used in describing the polarized emission of fluorophores and still commonly found in the literature is *polarization, P*:

$$P = (I_\parallel - I_\perp)/(I_\parallel + I_\perp) \tag{21}$$

The conversion from polarization to anisotropy is straightforward:

$$r = 2P/(3 - P) \tag{22}$$

Although the anisotropy function is perhaps more prevalent in the current literature, primarily because its use simplifies the derived mathematical expressions for complex cases containing several species, the information content of both functions is identical. The anisotropy of a sample containing "i" species is

$$r = \sum_i f_i r_i \tag{23}$$

where r_i and f_i are the emission anisotropy and the fractional contribution of the emission for the ith species, respectively. Similarly, the additivity of polarization (demonstrated by Weber, 1952) has the form:

$$\left(\frac{1}{P} - \frac{1}{3}\right)^{-1} = \sum_i f_i \left(\frac{1}{P_i} - \frac{1}{3}\right)^{-1} \tag{24}$$

thus, $1/P - 1/3$ is the harmonic mean of the individual polarization, P_i, weighted according to their fractional contributions to the total fluorescence.

4.3. Macromolecular Motion and Anisotropic Rotation

Anisotropy measurements report on a fluorophore's rotational diffusion and hence can provide information on structure and hydrodynamics of probe-associated macromolecules such as protein, RNA, and lipid aggregations. Consider a solution of fluorophore-labeled protein monomers which are spherical in shape and dissolved in an aqueous solution. At any given moment, the average orientation of the probe–protein conjugate is random. Each protein monomer rotates freely under the influence of thermal energy, i.e., Brownian motion without any external orienting force.

Upon excitation with polarized light, a fraction of the fluorophores are excited with a probability that is a function of the excitation intensity, the extinction coefficient, and θ, the angle between the axis of the polarized excitation and the axis of the absorption dipole. The absorption probability function with respect to θ is

$$f(\theta)\delta\theta = \cos\theta\sin\theta\delta\theta \tag{25}$$

The creation of a distribution of excited fluorophores, using light polarized along the z axis, is known as photoselection. With time, the excited fluorophores will return to a uniform random set. It is this relaxation process which is followed in order to determine rotational motion which, in turn, reflects the shape and size of the probe–protein conjugate. We observe this motion by measuring the state of emission anisotropy with time. The relationship between this time-dependent relaxation of the anisotropy and rotational diffusion is a sum of exponentials,

$$r(t) = r_0 \sum_i \alpha_i e^{-3t/\rho_i} \tag{26}$$

where α_i and ρ_i are the initial aniostropy amplitude (normalized such that $\sum_i \alpha_i = 1$) and the Debye rotational relaxation time for the ith rotational mode, respectively; t is time and r_0 is the limiting anisotropy. The rotational relaxation time, for the single exponential case (a sphere), can be described in terms of the hydrated molecular volume (V) of the rotating species, the viscosity (η), the temperature (T), and the gas constant (R):

$$\rho = 3V\eta/RT \tag{27}$$

The hydrated volume then can be related to the molecular weight of a protein through the specific volume, experimentally determined or calculated from the amino acid analysis (Cohn and Edsall, 1943; McMeekin *et al.*, 1949), and the hydration state which is typically in the range of 0.2–0.4 g H_2O/g protein.

We have, however, been assuming that the emission and absorption dipoles are colinear, a condition which is not always true (Weber, 1966). The relationship between the emission and absorption angle and the limiting anisotropy of our spherical protein, the anisotropy in the absence of Brownian motion (r_0), is given in Eq. (28). In substituting values for λ, we find r_0 at a maximum of 0.400 when the dipoles are colinear and at a minimum of -0.200 when λ is 90°. In analysis of time-resolved measurements, the limiting anisotropy is often treated as a variable and fit to the data with the relaxation times. However, independent knowledge of r_0 can assist in the data interpretation. This information is gained by observing the probe or probe–macromolecule conjugate in a highly viscous environment where there is no appreciable local or global motion. The observed steady-state anisotropy will be equivalent to the limiting anisotropy. Care must be exercised since the limiting anisotropy is dependent on the excitation wavelength and, though usually to a lesser extent, on the probe environment.

$$r_0 = (3\cos^2\lambda - 1)/5 \tag{28}$$

Some proteins do appear to behave as spherical bodies, but most are asymmetric and cannot, or should not, be treated as having a single rotational relaxation time (Fig. 6). In the case of a sphere there is a single rotational diffusion coefficient, D, inversely proportional to the rotational relax-

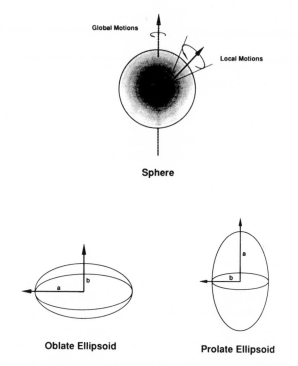

FIGURE 6. Ellipsoids of revolution commonly used as models of macromolecular shape. Rotational motions are defined around the major, a, and minor, b, ellipsoidal axes and the attachment point of the observed fluorophore.

ation time, ρ, and another commonly used value, the rotational correlation time, σ:

$$\rho = 3\sigma = 1/(2D) \tag{29}$$

However, when the more complex shape of a general ellipsoid is considered, the rotational motion is described by three rotational diffusion coefficients, one for each of the three unique axes. The relationship between the rotational relaxation times and rotational diffusion coefficients is no longer simple; five rotational relaxation and correlation times must be used to model the complex rotation of a general ellipsoid (Perrin, 1934; Belford *et al.*, 1972). Under the simplifying assumptions of prolate and oblate ellipsoids, where two of the three axes are equivalent, the mathematical expression reduces to three rotational correlation times. We give the simplified equation for anisotropy decay for symmetrical ellipsoids without rigorous mathematical treatment, which has been reviewed by others (Yguerabide, 1972; Belford *et al.*, 1972; Small *et al.*, 1988; Bucci and Steiner, 1988):

$$r(t) = \beta_1 e^{-t(4D_1 + 2D_2)} + \beta_2 e^{-t(D_1 + 5D_2)} + \beta_3 e^{-t(6D_2)} \qquad (30)$$

containing the rotational correlation times,

$$\sigma_1 = \frac{1}{(4D_1 + 2D_2)} \qquad \sigma_2 = \frac{1}{(D_1 + 5D_2)} \qquad \sigma_3 = \frac{1}{(6D_2)} \qquad (31)$$

where β_1, β_2, and β_3 represent complex expressions for the angles among the absorption and emission dipoles and the axes of the ellipsoid, t is time, r is the anisotropy, and D_1 and D_2 are the diffusion coefficients around the axis of symmetry and equatorial axes, respectively.

We have until now been considering probes, and probes bound to macromolecules, which rotate freely. There may be conditions, however, when this scenario does not hold, e.g., when a probe is restricted to motion described by some angle. In such a situation the anisotropy will not decay to zero (within a time not appreciably larger than several fluorescence lifetimes). If we consider a protein with a covalently attached probe, the observed anisotropy decay will derive from the global motion of the protein, the segmental motion of the protein domain that contains the fluorophore, and the local motion of the probe at its point of attachment (Fig. 6). We have already discussed global motions (*vide supra*), but the other two conditions possess restricted motion which alone would not be expected to show an anisotropy decay to zero (within a time not appreciably longer than several fluorescence lifetimes).

The distinction between local and structural motions is not always clear. Local motion is usually fast and in many cases approaches that of a probe free in solution while structural motion is expected to be slower since it involves the movement of larger protein domains. Practically, it is a difficult task to clearly distinguish between such motions which may range over a distribution of amplitudes.

There are two conditions of restrained rotators commonly encountered: restricted motion of a probe in the presence and in the absence of coupled rotations which are unrestricted. The former, case 1, would describe an oriented probe in an ordered bilayer or a probe bound to a macromolecule of infinite size, i.e., having a rotational relaxation time too slow to be observed on the fluorescence time scale. The latter condition would describe a probe wobbling at its attachment site covalently bound to a protein. Let us then examine the mathematical expressions for these two cases.

Case 1

In the presence of a single, restrained anisotropy decay, the anisotropy will not decay to zero but to a steady value, r_∞. It is generally assumed that the decay from the limiting anisotropy, r_0, to r_∞ will be described by a sum of exponentials. The expression then should contain a decay with limits of r_0 and r_∞ along with the constant term A_∞. Detailed description of rigorous mathematical treatment has been presented by Kinosita and co-workers (Kinosita *et al.*, 1977) who also give a suitable approximate expression as

$$r(t)/r_0 = A_\infty + (1 - A_\infty)e^{-D_w t/Z} \tag{32}$$

where

$$Z/D_w = \sigma_\omega \tag{33}$$

and

$$A_\infty = r_\infty/r_0 = \frac{1}{2}\cos\theta_{max}(1 + \cos\theta_{max})^2 \tag{34}$$

The rotational correlation time of the restricted local motion is σ which is restricted about a cone angle of θ_{max}. D_w is the wobbling diffusion constant and, here, z is an approximate relation determined from θ_{max}. Additional reviews and theoretical discussions on the details of these equations can be found in the literature (Wallach, 1967; Weber, 1978; Lipari and Szabo, 1980; Yguerabide and Yguerabide, 1984).

Case 2

Restrained anisotropy decay in which the global rotational rate cannot be neglected is a more complicated version of case 1 in which an additional rotational correlation time, not restricted, is present. An expression for this state can be derived from the approximate relationship of case 1 multiplied by a second exponential decay conveying the unrestricted motion (Lipari and Szabo, 1980). The relationship is thus

$$r(t)/r_0 = A_\infty e^{(-t/\sigma_1)} + (1 - A_\infty)e^{-t(1/\sigma_1 + 1/\sigma_2)} \tag{35}$$

where σ_1 and σ_2 are the global and local rotational correlation times, respectively. It is important to note that data collected on a case 2 system will fit to a two-component decay, one of which can be treated as the global rate, σ_1, but the other is a combination of both. When $\sigma_1 \gg \sigma_2$, then $(1/\sigma_1 + 1/\sigma_2)$ can be considered equal to $(1/\sigma_2)$. As σ_1 approaches infinity, Eq. (35) will become identical to Eq. (32).

4.4. Differential Phase and Modulation Fluorometry

4.4.1. Instrumentation

With the addition of polarizers to the excitation and emission ports, the equipment used for lifetime measurements can be used to measure anisotropy decay. The polarizers permit selection of properly polarized light for sample excitation, along the z axis (see Fig. 5), and discrimination of the parallel and perpendicular components of the emission. Similar to the analysis on lifetime decays, a range of modulation frequencies are employed to resolve fast and slow decays, in this case anisotropy decays, by calculating the difference between the phase angle and modulation ratio of the parallel and perpendicular components of the emission. Data analysis is then carried out for exponential decays, transforms of the equations given above, in the frequency domain.

Several practical points should be mentioned here. The first is that polarizers can vary in efficiency, ability to block light polarized along an incorrect axis, and optical density. Calcite prism polarizers, though expensive, are usually the polarizers of choice since they are highly efficient and have low absorption in the ultraviolet range above 240 nm. Film polarizers, in contrast, are less efficient and may absorb highly in the ultraviolet. The film variety do have the advantage of being considerably less expensive than the prisms (see Jameson *et al.,* 1978, for a comparison of several different types of polarizers). The orientation of the polarizers is also important since the analytical expressions for the data assume ideal conditions. One can easily check for polarizer alignment with a dilute solution of glycogen which, being a dipolar scatterer, should give a polarization close to 1.0. Weber (1956) has discussed in detail the influence of polarizer alignment and the numerical aperture of the light collection optics on the measured polarization.

In addition to artifacts caused by inefficient or poorly aligned polarizers, there are several other instrumental problems that can lead to trouble. Instruments often have a bias in the measurement of polarized light and will report lower or higher intensities for one component leading to inaccurate polarizations. To correct for this, one excites a fluorescent solution with light polarized along the y axis which creates a condition where $I_x = I_z$ and therefore $I_\perp = I_\parallel$, despite the sample anisotropy. The correction factor, $I_\perp / I_\parallel$, must be used to adjust the intensities to their proper level. The correction factor should be measured under the expected experimental conditions since its value is dependent on wavelength and monochromator band-pass. When filters are used in place of a monochromator, the correction is minimized but in these cases too, the correction factor should be measured. Movement of the emission polarizer, necessary to measure both emission components, can change the focus plane of the fluorescence on the photomultiplier tube and result in polarization artifacts.

4.4.2. Data Analysis and Examples

The expressions for differential phase and modulation for a single, spherical rotator are

$$\Delta\phi = \tan^{-1}\left[\frac{18\omega r_0 R}{(k^2 + \omega^2)(1 + r_0 - 2r_0^2) + 6R(6R + 2k + kr_0)}\right] \quad (36)$$

$$Y^2 = \frac{[(1 - r_0)k + 6R]^2 + (1 - r_0)^2\omega^2}{[(1 + 2r_0)k + 6R]^2 + (1 + 2r_0)^2\omega^2} \quad (37)$$

where $\Delta\phi$ is the phase difference, Y the modulation ratio ($\perp/\parallel$) of the AC components, ω the angular modulation frequency, r_0 the limiting anisotropy, k the radiative rate constant ($1/\tau$), and R the rotational diffusion coefficient (Weber, 1977). The values measured are the modulation and the phase difference which are then fit to a single exponential model above. Using a least-squares routine over the frequency range, the minimizing procedure follows the goodness of the fit to a reduced chi-square (Bevington, 1969).

Differential phase data plotted as a function of modulation frequency for a simple rotator show a smooth bell-shaped curve having a peak magnitude and peak frequency determined by the rotational rate, fluorescence lifetime, and the limiting anisotropy. Illustrated in Fig. 7 are the expected differential phase curves for similarly labeled species with Debye rotational relaxation times of 3, 30, and 300 nsec. As the relaxation time is increased, and the motions become slower, the differential phase peak shifts to lower frequencies and decreases in amplitude. For more complex systems, the appearance of multiple peaks and shoulders become important characteris-

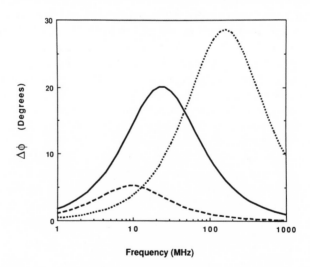

FIGURE 7. Differential phase data for an isotropic rotator with a 3-nsec (dotted line), 30-nsec (solid line), or 300-nsec (dashed line) rotational relaxation time. In each case a lifetime of 20 nsec was used and colinear excitation and emission dipoles were assumed.

tics as well. The expression for a single rotational relaxation time has been given above [Eq. (36)].

In addition to the phase data, modern multifrequency phase fluorometers also provide the modulation ratio, at each frequency, of the parallel and perpendicular emission components. From the modulation data one can derive the anisotropy amplitude and rotational relaxation times [Eq. (37)] independently. The modulation and differential phase data weight the information differently and, for that reason, provide complementary data on the sample rotational rates. Curves of the expected modulation ratio with respect to frequency are drawn in Fig. 9; the rotational relaxation rates and lifetimes are identical to those used in Fig. 8. A region of demodulation, characteristic of the rotational parameters, is the normal shape of the modulation curves. The influence of increasing rotational relaxation times on the modulation profile (Fig. 8) is to reduce the modulation ratio at low frequencies and shift the regions of increased demodulation toward lower frequencies. Though not as visually descriptive as the differential phase data, the modulation ratio is equally sensitive to the sample rotational parameters. Both data types are commonly collected and analyzed simultaneously to increase the accuracy and precision of the recovered rotational parameters.

Data sets for two samples are given in Fig. 9 where the characteristic patterns for single relaxation times are clearly visible. In this example, the differential phase and modulation data for free and tRNA-bound ethidium bromide are given. Bound to tRNA (at saturating tRNA concentrations) the

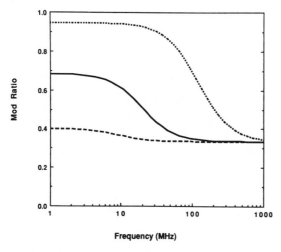

FIGURE 8. Modulation ratio as a function of frequency for the species given in Fig. 7.

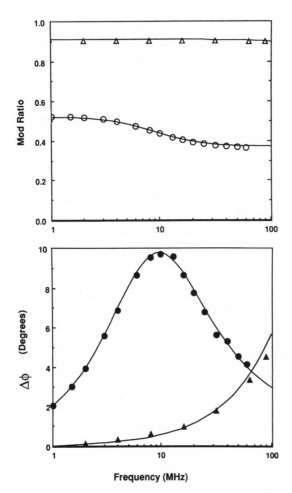

FIGURE 9. Differential phase (closed symbols) and modulation (open symbols) data for ethidium bromide in solution (triangles) and ethidium bromide bound to tRNA (circles). The resolved rotational relaxation times (5°C) were 144 and 0.5 nsec for free and bound ethidium bromide, respectively. The curves are the least-squares fit to the data.

ethidium bromide has a rotational relaxation time of 144 nsec reflecting the macromolecular motion of the tRNA molecule. The motion of free probe, however, is much faster causing the differential phase peak and demodulation to shift to higher frequencies, greater than the given data range. Thus, for free ethidium bromide one sees the initial increase of the expected differential phase "bell-shaped" curve and the low-frequency plateau of the modulation data. It should, as a practical point, be mentioned that the frequency range will impact on the precision of the results; in general, the more complete the bell-shaped curve, the higher the precision. Limitations in the equipment and low sample signal can, however, restrict the usable frequency range. Though the tRNA–ethidium bromide system is actually more complex than a single exponential decay (see Hazlett and Jameson, 1988, and Hazlett *et al.*, 1989, and references therein), the presented data display quite well the expected curves for a simple system.

As a rule, most systems cannot be described by a single exponential decay; many have multiple species each with one or more associated rotational relaxation times. One of the most common exponential questions related to a complex system is whether a resolved fast motion is due to probe movement at its binding site or the presence of free probe. Clearly, a single species with two rotational relaxation times is physically distinct from two species each with its own rotational relaxation time. In Fig. 10 (from Jameson *et al.*, 1987) two sets of phase and modulation data illustrate the former case. The data were taken from the fluorescence emission of the single tryptophan of *Escherichia coli* elongation factor Tu (EF-Tu) in the presence and absence of elongation factor Ts (EF-Ts) which binds EF-Tu and has no contributing tryptophan fluorescence. Analysis of the results gave global rotational relaxation times of 63 and 84 nsec for the EF-Tu and EF-Ts · EF-Tu, respectively; reasonable for the size and shape of the complexes. Examining the high-frequency region of the differential phase data, one notes in both data sets an increase indicating the presence of a fast motion. Furthermore, upon the formation of the EF-Ts · EF-Tu complex, a pronounced increase in phase difference at the high frequencies is observed. The interpretation of this result is that the EF-Tu region containing Trp-184 is loosened in complex formation, permitting an increase in the rotational speed and amplitude of the tryptophan residue. Here, the data were analyzed without consideration of the hindered aspect of the local motion. In many cases the analysis does not reveal this hindered motion unless the local motion is a significant part of the total anisotropy amplitude. Discussion on the use of the phase

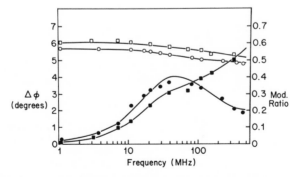

FIGURE 10. Multifrequency differential phase (closed symbols) and modulation (open symbols) data for elongation factor Tu complexed with GDP (circles) and elongation factor Ts (squares). Curves represent the least-squares fit to the data.

and modulation methods for hindered motion has been offered by others (Weber, 1978; Lakowicz *et al.,* 1985).

When two or more rotating species are present, analysis must consist of multiple rotational relaxation times, but, as we described above, this is also true of the single species, multiple rotational relaxation time situations. If free probe were present in the example above, we would have one species with two rotational relaxation times, one a free probe with a single relaxation time. One differentiates free probe from local motion by either rigidly establishing a hindered local motion (not easily accomplished) or observing characteristic phase delays caused by differences in the lifetimes of free and bound probe. The effect of lifetime differences in conjunction with rotational relaxation time differences can have a profound influence on the differential phase and modulation results; permitting increased resolution of the decay components. The association of specific lifetime of a heterogeneous system to a specific rotational relaxation time of a multiple species system is known as associative decay, first noted as anomalous phase delay by Weber (Spencer and Weber, 1970) and recently discussed by others (Brand *et al.,* 1985; Szmacinski *et al.,* 1987; Hazlett and Jameson, 1988). Stressed in our work (Hazlett and Jameson, 1988) is that this effect can be used to identify and quantify the presence of small amounts of free (or bound) species in a mixture. It should be noted that lifetime analysis alone is not determining since the presence of heterogeneous decay in a bound fluorescence probe is common.

Figure 11 ideally illustrates associative decay. Both free ethidium bro-

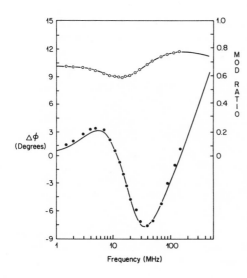

FIGURE 11. An example of anomalous phase delay resulting from associative decay. Differential phase (●) and modulation (○) data and least-squares fit for a two-component system containing free ethidium bromide and ethidium bromide bound to tRNA are shown.

mide and tRNA-bound ethidium bromide are present and result in the observed negative phase delay, a counterintuitive effect that is, however, predicted by theory. The modulation data are equally affected by the associative decay showing an increase, instead of a decrease or leveling, at frequencies > 20 mHz (Fig. 11). The primary condition in which a negative phase delay such as this will be observed is when the free probe has a faster rotation and shorter lifetime than the bound probe (Hazlett and Jameson, 1988). Both conditions are common for covalent and noncovalent association of fluorophores to proteins. However, free and bound probe with identical lifetimes will not yield anomalous phase and modulation results; analysis must then be performed to reveal or exclude the presence of a hindered motion. One possible method to artificially attain a lifetime difference is to include a soluble quencher which would lower the free probe lifetime with respect to the lifetime of the often more protected bound probe. The presence of associative decay would then indicate free probe.

5. PHASE-RESOLVED METHODS

The technique of phase-sensitive detection has had wide-spread applicability in spectroscopy for many years. In fact, one of the more celebrated examples of phase-sensitive detection was the original experiment by Fizeau (1849) who was able to realize terrestrial measurements of the speed of light by observing the occlusion of light through a rotating toothed wheel, i.e., a phase-sensitive detector system (Fizeau, 1849). One of the first applications of the method in fluorescence, however, was by Veselova *et al.* (1970), who used the technique to resolve the individual spectrum associated with components in a mixture of fluorophores. In the simplest sense, phase-sensitive detection is a simple form of time discrimination. The basic operational principle behind phase-resolved fluorometry is that excitation of a population of fluorophores, composed of species which emit with characteristically different lifetimes and spectra, by a source with a sinusoidally modulated intensity permits spectral resolution of the different components through the use of detectors sensitive to the phase angle. Advances in this area have been made in both hardware and software. The approaches based on phase-sensitive electronics have been reviewed (McGown and Bright, 1984). More recently, however, a software-based approach was implemented which offers advantages over the hardware approach in particular cases (Gratton and Jameson, 1985). The mathematical basis for phase-resolved fluorescence has been described in detail (Gratton and Jameson, 1985; Barbieri *et al.,* 1988). Basically, Eqs. (7)–(11) are utilized to calculate the values of f_i as a function of wavelength which yield a lifetime resolved spectrum. The explicit expression for the fractional contribution calculated from the phase data is:

$$R_P = \frac{\tau_2 - \tau_P}{\tau_P - \tau_1} \cdot \frac{1 + \omega^2\tau_1^2}{1 + \omega^2\tau_2^2} \tag{38}$$

where $R = f_1/f_2$. The fractional intensities may be converted into molar ratios by taking into account the relative yields of the two species.

A typical application of the phase-resolved procedure is shown in Fig. 12 (from Gratton and Jameson, 1985). The intensity spectrum in this figure was obtained from a mixture of DENS, POPOP, and perylene in ethanol. The emission spectra corresponding to the individual components are shown in the figure inset. The phase-resolved spectra were obtained using 20-MHz modulation frequency and component lifetimes of 10.8 nsec (DENS), 4.3 nsec (perylene), and 0.9 nsec (POPOP). The phase-resolved method has also been recently utilized to resolve porphyrin–photoproduct mixtures (Valat et al., 1988).

This procedure can also be utilized to separate fluorescence from phosphorescence contributions. Figure 13 shows the phase-resolved spectra of erythrosin in glycerol (from Barbieri et al., 1988). The phosphorescence contribution is weak relative to the fluorescence but can be isolated using the phase-resolved method. In this case the fluorescence lifetime was 100 psec while the phosphorescence lifetime was on the order of 10 nsec. The modulation frequency used for this determination was 300 kHz. In these cases where

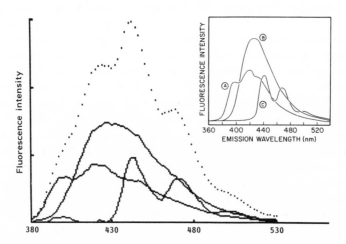

FIGURE 12. Emission spectra of the individual components. (Inset) POPOP (A), DENS (B), and perylene (C) in ethanol (not degassed). Intensity (dotted line) and phase-resolved spectra (solid lines) for the ternary mixture.

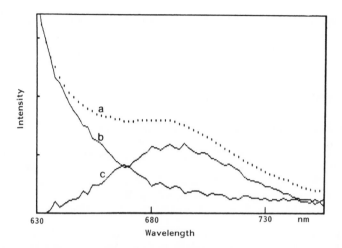

FIGURE 13. The total intensity (a) and the phase-resolved fluorescence (b) and phosphorescence (c) spectra of erythrosin in glycerol.

the lifetimes are very different, one finds that the spectral separation achieved is not particularly dependent upon the exact values used in the phase-resolution analysis.

The software approach to phase-resolved spectra is, in fact, only a special case of the more complete global analysis of the wavelength-dependent decay data. Yet, in many cases where some information is already available, such as the lifetime of one or more components, this approach can be highly expeditious.

6. FUTURE DIRECTIONS

The future development of phase and modulation fluorometry in the biological sciences will almost surely proceed in several directions. First, the useful frequency range will be extended to both lower and higher limits. The lower modulation frequencies will be applied to phosphorescence or delayed fluorescence probes and will be suitable for studying slow rotational motions, such as those expected for membrane-associated proteins. The parallel frequency methods will develop so that data from a number of frequencies will be obtained simultaneously using pulsed excitation and FFT-type analysis. Finally, but perhaps most significantly, all these approaches will eventually be applied through the microscope to bring the full power of the technique to bear on living cells. When this step is realized, then fluorescence microscopy will truly be quantitative.

ACKNOWLEDGMENTS. Some of the work presented here was supported by National Science Foundation Grants DMB 8706440 and DMB 9005195. D.M.J. is an Established Investigator of the American Heart Association.

7. REFERENCES

Alcala, J. R., Gratton, E., and Jameson, D. M., 1985, *Anal. Instrum.* **14**:225–250.

Alcala, J. R., Gratton, E., and Prendergast, F. G., 1987, *Biophys. J.* **51**:587–596.

Badea, M. G., and Brand, L., 1979, *Methods Enzymol.* **61**:378–425.

Barbieri, B. F., DePiccoli, F., and Gratton, E., 1988, *Proc. S.P.I.E.* **909**:366–369.

Beechem, J. M., and Brand, L., 1985, *Annu. Rev. Biochem.* **54**:43–71.

Beechem, J. M., and Gratton, E., 1988, *Proc. S.P.I.E.* **909**:70–81.

Beechem, J. M., Knutson, J. R., Ross, J. B. A., Turner, B. W., and Brand, L., 1983, *Biochemistry* **22**:6054–6058.

Beechem, J. M., Knutson, J. R., and Brand, J. R., 1986, *Biochem. Soc. Trans.* **14**:832–835.

Belford, C. G., Belford, R. L., and Weber, G., 1972, *Proc. Natl. Acad. Sci. USA* **69**:1392–1393.

Bevington, P. R. (ed.), 1969, *Data Reduction and Error Analysis in the Physical Sciences,* McGraw-Hill, New York.

Birks, J. B., and Munro, I. H., 1967, *Prog. React. Kinet.* **4**:239–303.

Brand, L., Knutson, J. R., Davenport, L., Beechem, J. M., Dale, R. E., Walbridge, D. G., and Kowalczyk, A. A., 1985, in *Spectroscopy and the Dynamics of Molecular Biological Systems,* P. M. Bayley and R. C. Dole eds. Academic Press, London.

Bucci, E., and Steiner, R. F., 1988, *Biophys. Chem.* **30**:199–224.

Cohn, E. L., and Edsall, J. T., 1943, in *Proteins, Amino Acids and Peptides as Ions and Dipolar Ions,* Reinhold, New York.

Cundall, R. B., and Dale, R. E. (eds.), 1983, *Time-Resolved Fluorescence Spectroscopy in Biochemistry and Biology,* Plenum Press, New York.

Demas, J. N., 1983, *Excited State Lifetime Measurements,* Academic Press, New York.

Duschinsky, F., 1933, *Z. Phys.* **81**:7–22.

Eftink, M. R., and Wasylewski, Z., 1989, *Biochemistry* **28**:382–391.

Fiorini, R. M., Valentino, M., Glaser, M., Gratton, E., and Curatola, G., 1988, *Biochim. Biophys. Acta* **939**:482–485.

Fizeau, A., 1849, *C.R. Acad. Sci.* **31**:90–93.

Gaviola, E., 1926, *Ann. Phys. (Leipzig)* **81**:681–710.

Gratton, E., and Barbieri, B., 1986, *Spectroscopy* **1**:28–36.

Gratton, E., and Jameson, D. M., 1985, *Anal. Chem.* **57**:1694–1697.

Gratton, E., and Limkeman, M., 1983, *Biophys. J.* **44**:315–324.

Gratton, E., Jameson, D. M., and Hall, R. D., 1984a, *Annu. Rev. Biophys. Bioeng.* **13**:105–124.

Gratton, E., Jameson, D. M., Rosato, N., and Weber, G., 1984b, *Rev. Sci. Instrum.* **55**:486–494.

Hazlett, T. L., and Jameson, D. M., 1988, *Proc. S.P.I.E.* **909**:412–419.

Hazlett, T. L., Johnson, A. E., and Jameson, D. M., 1989, *Biochemistry* **28**:4109–4117.

Jameson, D. M., and Gratton, E., 1983, in *New Directions in Molecular Luminescence* (D. Eastwood, ed.), ASTM STP 4822, American Society for Testing and Materials, Philadelphia, p. 67.

Jameson, D. M., and Reinhart, G. D. (eds.), 1989, *Fluorescent Biomolecules: Methodologies and Applications,* Plenum Press, New York.

Jameson, D. M., Weber, G., Spencer, R. D., and Mitchell, G., 1978, *Rev. Sci. Instrum.* **49**:510–514.

Jameson, D. M., Gratton, E., and Hall, R. D., 1984, *Appl. Spectrosc. Rev.* **20**:55–106.
Jameson, D. M., Gratton, E., and Eccleston, J. F., 1987, *Biochemistry* **26**:3894–3901.
Kinosita, K., Jr., Kawato, S., and Ikegami, A., 1977, *Biophys. J.* **20**:289–305.
Knutson, J. R., Beechem, J. M., and Brand, L., 1983, *Chem. Phys. Lett.* **102**:501–507.
Lakowicz, J. R., 1983, *Principles of Fluorescence Spectroscopy,* Plenum Press, New York.
Lakowicz, J. R., Cherek, H., and Maliwal, B. P., 1985, *Biochemistry* **24**:376–383.
Laws, W. R., and Jameson, D. M., 1989, in *Synchrotron Radiation in Structural Biology* (R. M. Sweet, ed.), Plenum Press, New York.
Lipari, G., and Szabo, A., 1980, *Biophys. J.* **30**:489–506.
McGown, L. B., and Bright, F. V., 1984, *Anal. Chem.* **56**:2195–2199.
McMeekin, T. L., Groves, M. L., and Hipp, N. J., 1949, *J. Am. Chem. Soc.* **7**:3298–3300.
Neal, S. E., Eccleston, J. F., Hall, A., and Webb, M. R., 1988, *J. Biol. Chem.* **263**:19718–19722.
Nordlund, T. M., 1988, *Proc. S.P.I.E.* **909**:35–50.
Perrin, F., 1934, *J. Phys. Radium* **5**:487–511.
Small, E. W., Libertini, L. J., and Small, J. R., 1988, *Proc. S.P.I.E.* **909**:97–107.
Spencer, R. D., and Weber, G., 1969, *Ann. N.Y. Acad. Sci.* **158**:361–376.
Spencer, R. D., and Weber, G., 1970, *J. Chem. Phys.* **52**:1654–1663.
Szmacinski, H., Jayaweera, R., Cherek, H., and Lakowicz, J. R., 1987, *Biophys. Chem.* **27**:233–241.
Taylor, D. L., Waggoner, A. S., Lanni, F., Murphy, R. F., and Birge, R. R. (eds.), 1987, *Applications of Fluorescence in the Biomedical Sciences,* Liss, New York.
Teale, F. W. J., 1983, in *Time-Resolved Fluorescence Spectroscopy in Biochemistry and Biology* (R. B. Cundall and R. E. Dale, eds.), Plenum Press, New York, p. 59.
Valat, P., Reinhart, G. D., and Jameson, D. M., 1988, *Photochem. Photobiol.* **47**:787–790.
Veselova, T. V., Limareva, L. A., and Cherkasov, A. S., 1965, *Izv. Akad. Nauk SSSR Ser. Fiz.* **29**:1340–1345.
Veselova, T. V., Cherkasov, A. S., and Shirokov, A. S., 1970, *Opt. Spectrosc.* **29**:617–618.
Wallach, D., 1967, *J. Chem. Phys.* **47**:5258–5268.
Weber, G., 1952, *Biochem. J.* **51**:145–155.
Weber, G., 1966, in *Fluorescence and Phosphorescence Analysis* (D. M. Hercules, ed.), Interscience, New York, pp. 217–240.
Weber, G., 1956, *J. Opt. Sci. Amer.* **46**:646–655.
Weber, G., 1977, *J. Chem. Phys.* **66**:4081–4091.
Weber, G., 1978, *Acta Phys. Pol. A* **54**:859–865.
Weber, G., 1981, *J. Phys. Chem.* **85**:949–953.
Williams, B. W., and Stubbs, C. D., 1988, *Biochemistry* **27**:7994–7999.
Yguerabide, J., 1972, *Methods Enzymol.* **26**:498–578.
Yguerabide, J., and Yguerabide, E. E., 1984, in *Optical Techniques in Biological Research* (D. L. Rousseau, ed.), Academic Press, New York.

Chapter 5

Fluorescence Investigations of Receptor-Mediated Processes

William J. Phillips and Richard A. Cerione

1. INTRODUCTION

1.1. Examples of Receptor-Coupled Signaling Pathways

1.1.1. Regulation of Adenylate Cyclase Activity and Vertebrate Vision

There are a number of receptor-coupled signal transduction systems which regulate the levels of second messengers within the cell through the actions of at least three types of protein components. Specifically, these components are the receptor protein itself, a GTP-binding protein (G protein), and a biological effector which can be an enzyme or an ion channel (Gilman, 1987; Cerione et al., 1986; Neer and Clapham, 1988; Stryer et al., 1981). The G proteins function as transducers within these systems by mediating the regulation of effector activities by specific cell surface receptors. Among the types of biological activities which are regulated in this manner include hormonal-regulated adenylate cyclase activity (Gilman, 1987; Cerione et al., 1986; Lefkowitz and Caron, 1988), vertebrate vision (Stryer et al., 1981; Fung, 1983), receptor-coupled phosphoinositide metabolism (Berridge and Irvine, 1984; Gomperts, 1983; Wallace and Fain, 1985; Bokoch and Gilman, 1984; Smith et al., 1986), arachidonic acid release (Burch et al., 1986; Jelsema and Axelrod, 1987), activation of Ca^{2+} channels found in excitatory cells (Hes-

WILLIAM J. PHILLIPS and RICHARD A. CERIONE • Department of Pharmacology, Cornell University, Ithaca, New York 14853-6401.

cheler *et al.,* 1987; Yatani *et al.,* 1987), and cardiac K^+ channels (Codina *et al.,* 1987; Logothetis *et al.,* 1987). Various lines of evidence suggest that a common regulatory mechanism underlies the functioning of each of these signaling pathways. In all cases, the signal is initiated by the binding of an appropriate ligand (i.e., hormone) to the receptor, or in the case of visual transduction, by the absorption of light. This constitutes the activation of the receptor, thereby promoting the interaction of the receptor with the G protein. It is the formation of the receptor–G protein complex which is responsible for the exchange of a tightly bound molecule of GDP for GTP. The binding of GTP (or a nonhydrolyzable GTP analogue) then primes the G protein for interacting with the effector protein. This interaction is responsible for the regulation of the effector activity with the regulation persisting until the bound GTP is hydrolyzed to GDP. The GTPase activity results in the deactivation of the G protein and thereby provides the means by which the signaling system can return to its initial (resting) state.

In two of the receptor-coupled signaling systems described above, i.e., the β-adrenergic receptor-coupled adenylate cyclase system and the rhodopsin-coupled phototransduction system, the primary components participating in the signaling pathways have been well characterized. The mammalian β-adrenergic receptor (βAR) is a single-chain, transmembranal glycoprotein of M_r 65,000 (Benovic *et al.,* 1984). Similarly, rhodopsin is a single-chain glycoprotein of M_r 40,000 (Hargrave *et al.,* 1983). Both of these receptors contain appreciable hydrophilic, extracellular, and cytoplasmic domains and are comprised of seven transmembranal helical segments (Hargrave *et al.,* 1983; Dixon *et al.,* 1986). The latter appear to contain the sites for hormone binding in the case of βAR, and for the covalent attachment of the chromophore, retinal, in the case of rhodopsin. It is worth noting that a number of other receptor proteins which are known, or suspected, to couple to G proteins also contain similar structural features; these include the α_2-adrenergic receptor, which is involved in mediating the inhibition of adenylate cyclase activity (Kobilka *et al.,* 1987), and the muscarinic (M_{1-4}) acetylcholine receptors (Kubo *et al.,* 1986a,b) which participate in the inhibition of adenylate cyclase and in the stimulation of phosphoinositide metabolism.

As alluded to above, the G proteins represent the cornerstone of these signaling systems since they shuttle information between the receptor proteins and the biological effectors. It is the activation and deactivation of these proteins which has received the greatest attention, thus far, in terms of the application of fluorescence approaches (see Section 2). Like the case for the receptor proteins, the G proteins participating in the stimulation of adenylate cyclase activity (designated as G_s) and in the rhodopsin-coupled stimulation of the retinal cyclic GMP phosphodiesterase (the retinal G protein being designated as transducin) also share a great deal of structural homology (Gilman, 1987). These transducers are comprised of three types of subunits which are designated as α, β, and γ. The α subunits appear to be structurally distinct in each of the well-characterized members of the G protein family, ranging in size from 45–52,000 for the G_s protein to 39,000 for transducin. This subunit is responsible for most of the functional activities of the G protein, including GTP binding, GTPase activity, and in the majority of cases, for coupling to the specific biological effectors. Thus, it is this subunit in particular which is of interest in structure–function studies using fluorescence spectroscopy (see below). All of the well-characterized G proteins appear to share a structurally similar, if not identical, β subunit (M_r 35–36,000). Likewise, G_s, and the inhibitory G proteins of the adenylate cyclase system (designated as G_i), as well as the bovine brain G_o protein (Hildebrandt et al., 1985; Cerione et al., 1987), all appear to contain identical low-molecular-weight γ subunits (M_r 5–10,000). However, the γ subunit of transducin appears to be structurally distinct from those of the other members of the family.

The structure–function similarities which are shared among the different biological effectors remain to be determined, although, thus far, it appears that the effector enzymes operating in the adenylate cyclase and phototransduction systems possess quite different structures. Specifically, the cyclic GMP phosphodiesterase (PDE) of the vertebrate vision system is made up of two large subunits, designated as α and β (M_r 85–88,000), and a smaller subunit designated as γ (M_r 14,000) (Baehr et al., 1979; Kohnken et al., 1981), while adenylate cyclase is suspected to be comprised of a single type of polypeptide of M_r 130–150,000 (Pfeuffer et al., 1983; Smigel, 1986). Nonetheless, a common structural domain which is responsible for binding the activated G protein is likely to be present on these different effectors.

1.1.2. Growth Factor Receptor/Tyrosine Kinases

The mechanisms by which polypeptide growth factors trigger a mitogenic response are certain to involve multicomponent signaling pathways. Various lines of evidence, in fact, suggest that GTP-binding proteins might serve as transducers in growth factor-coupled signal transduction (Johnson et al., 1986; Heyworth and Houselay, 1983; Heyworth et al., 1985; Gawler and Houselay, 1987; Luttrell et al., 1988). Two of the best studied members of the growth factor receptor/tyrosine kinase family are the epidermal growth factor (EGF) receptor and the insulin receptor. Primary structural information has been obtained for both of these receptors based on cDNA cloning efforts. The EGF receptor appears to be comprised of a single type of polypeptide, M_r 170,000 (Ullrich et al., 1984), while the insulin receptor is comprised of two types of polypeptides, designated α (M_r 170,000) and β (M_r

95,000), which are connected by disulfide bridges to form an α_2–β_2 arrangement (Ullrich *et al.*, 1985). Each α–β pair of the insulin receptor, as well as the monomeric EGF receptor, is comprised of two major hydrophilic domains which are linked by a very short (23 amino acids) transmembranal region. The growth factor binding sites are located within the hydrophilic region which is accessible to the extracellular space (i.e., the α subunit of the insulin receptor) while the tyrosine kinase domain is located on the opposite side of the membrane adjacent to the cytoplasm (i.e., the β subunit of the insulin receptor). There are a variety of effector activities which appear to be regulated by these growth factor receptor/tyrosine kinases including phosphoinositide metabolism and ion fluxes across the membranes. Thus, while the overall structures of these growth factor receptor/tyrosine kinases are significantly different from those of the βAR/rhodopsin family, it nonetheless is tempting to speculate that growth factor action involves multicomponent signaling systems where a transducer protein shuttles between these receptor/kinases and phospholipase C-like enzymes or ion channels.

1.2. The Use of Purification and Reconstitution Approaches to Characterize the Molecular Mechanisms Underlying Receptor-Coupled Signaling Pathways

A prerequisite for the structure–function studies of these different types of receptor-coupled signaling systems is the development of methods for reconstituting the interactions occurring between the signaling components in a well-defined, detergent-free medium (ideally in a lipid milieu as provided by a phospholipid vesicle). Over the past few years a great deal of success has been achieved in the development of reconstituted systems containing the purified components of the βAR-coupled adenylate cyclase system. Two-component βAR/G_s systems have been constructed which have been used to study each of the key interactions that occurs between the pure receptor and the pure G protein: these being a G_s-induced high-affinity binding by β-adrenergic agonists to βAR, an agonist-stimulated high-affinity binding of guanine nucleotides to G_s, and an agonist-stimulated GTPase activity in the G_s protein (Cerione *et al.*, 1984a, 1985; Brandt *et al.*, 1983; Asano *et al.*, 1984; Asano and Ross, 1984). The development of these two-component receptor/G protein systems was essential for determining the types of functional activities which can be attributed to pure receptor and G protein molecules. These reconstituted systems were then expanded to contain the effector enzyme, i.e., partially purified preparations of adenylate cyclase (C) (Cerione *et al.*, 1984b; May *et al.*, 1984). Fully reconstituted βAR/G_s/C systems demonstrated hormone-stimulated cyclic AMP production which was blocked by β-adrenergic antagonists like alprenolol or propranolol, with the appropriate stereoselectivity, and was inhibited in a GTPγS-dependent manner by the purified G_i protein.

The construction of these reconstituted systems seemed to set the stage for detailed studies of the molecular mechanisms responsible for receptor–G protein and G protein–effector interactions. An obvious long-term goal was to use spectroscopic (fluorescence) approaches to directly monitor each of the key steps underlying receptor–G protein coupling, the actual activation and deactivation events of the G protein, and G protein–effector interactions. However, various factors have limited the full exploitation of these reconstituted systems in examining such issues. Specifically, the amounts of βAR which can be routinely purified make it difficult to consistently construct phospholipid vesicle systems which contain >1 mole receptor per mole phospholipid vesicles. The stoichiometric incorporation of the receptor into lipid vesicles is desirable when studying receptor–G protein interactions since maximal effects will occur when all vesicles contain greater than 1 receptor and 1 G protein per lipid vesicle, as opposed to a situation where some vesicles contain only receptor molecules and others contain only G proteins. A second problem is related to the fact that all of the components of the signaling system—the receptor (βAR), the G protein (G_s), and the effector (adenylate cyclase)—are solubilized and purified in detergents. Thus, the construction of hormone-responsive reconstituted systems entails the insertion of each of the individual components into lipid vesicles (i.e., the detergents must be removed from each of the signaling components). Each of these proteins is typically inserted with a scrambled orientation so that roughly half of the G proteins and the effector enzymes are facing the inside of the vesicle and are inert to receptor-promoted events that would be initiated by post-adding a hormone to the outside of the lipid vesicle. Finally, a third problem concerns the overall degree of difficulty in purifying the adenylate cyclase enzyme in quantities approaching the levels of βAR and G_s, coupled to the fact that a great deal still remains to be determined about the structure–function properties of this enzyme.

For these reasons, we have turned our attention to the vertebrate phototransduction system. This system provides an excellent model for probing the detailed mechanisms of receptor/G protein-coupled signal transduction. Each of the primary components of this system can be obtained in milligram quantities. In addition, with the exception of rhodopsin, all of the components of this signaling pathway can be prepared and stored in the absence of detergent. Thus, different combinations of the retinal G protein, transducin, or the effector enzyme, the cyclic GMP PDE, can be added to a single preparation of rhodopsin-containing lipid vesicles (Cerione *et al.*, 1988). An added advantage of this system is the ready availability of rhodopsin (in milligram quantities). This enables the construction of lipid vesicles that contain many more than one rhodopsin molecule per vesicle, thus enhancing the probability of monitoring the functional interactions between receptors and G proteins.

In Section 2 of this chapter, we describe our use of the retinal visual system to develop fluorescence approaches for monitoring the receptor-stimulated activation–deactivation cycle of a G protein. The expectation is that these approaches will provide new insight into the molecular mechanisms underlying G protein function. In Section 3, we describe the use of fluorescence spectroscopy in addressing structure–function issues in growth factor-coupled signal transduction where the EGF receptor is being used as a model. Clearly, this receptor system lacks many of the advantages of the rhodopsin/transducin-coupled visual system. However, it does possess a distinct advantage regarding the ability to specifically label the growth factor molecule (EGF) with fluorophores. These fluorescent growth factors can then be used to monitor potential receptor-receptor interactions under a variety of conditions using fluorescence resonance energy transfer approaches. The ability to directly visualize these interactions should enable an unambiguous assessment of their roles in growth factor action.

1.3. Fluorescence Spectroscopy as a Tool for Studying Protein Structure–Function

1.3.1. Use of Intrinsic Protein Fluorescence as a Monitor for Protein–Ligand Interactions

Before we consider the use of fluorescence approaches in receptor-coupled multicomponent signaling systems, it is worthwhile to briefly review how fluorescence spectroscopy has been used in the past to study protein–ligand and protein–protein interactions. It has been well established that the intrinsic fluorescence of a protein typically reflects the emission of tryptophan residues, over a range of 315–350 nm, upon the excitation of the protein at 275–295 nm. Although tyrosine residues also are capable of fluorescence emission (300–310 nm, excitation 270–280 nm), in most cases this emission is buried within the tryptophan spectra due to the greater quantum yields of the tryptophan residues. Studies with model (tryptophan-containing) compounds have shown that the tryptophan emission is sensitive to the local environment, i.e., it can be affected by the presence of a neighboring charged residue or by the overall hydrophobicity of the environment (Brand and Witholt, 1967). In general, if a tryptophan residue is buried within a hydrophobic environment, its emission spectrum will be blueshifted relative to the spectrum for a solvent-accessible residue and the overall intensity of the fluorescence will be enhanced. Thus, the intrinsic fluorescence of a protein can be a powerful tool for studying ligand-induced conformational changes within the protein. One particular advantage is that it bypasses those difficulties which might arise regarding the maintenance of biological function, which is a factor that must be considered when chemically attaching fluorescent reporter groups to proteins (see below).

In Section 2 we describe how the intrinsic tryptophan fluorescence of G proteins has been used to probe various aspects of the activation and deactivation of these transducers. There is a good deal of precedent for the use of such strategies in studying the interactions of ligands or substrates with soluble proteins and enzymes. One relevant example has been the use of intrinsic fluorescence to study coenzyme binding to horse liver alcohol dehydrogenase. These approaches have allowed the detection of distinct conformational changes in the enzyme, which are elicited by different forms of the coenzyme (NAD^+, NADH), in a manner analogous to the detection of specific conformational changes in G proteins that are elicited by different forms of guanine nucleotides (see below). Horse liver alcohol dehydrogenase is a dimeric enzyme (M_r 160,000) which is involved in the conversion of ethanol to acetaldehyde; this reaction is accompanied by the conversion of the oxidized form of the coenzyme, NAD^+, to the reduced form, NADH. A good deal of structural information is known about the enzyme and in fact crystal structures of the enzyme and its various binary (enzyme–coenzyme) and ternary (enzyme–coenzyme–substrate) complexes have been determined (Branden *et al.,* 1975; Samana *et al.,* 1977). A number of studies have shown that the intrinsic tryptophan fluorescence of the enzyme is quenched at alkaline pH due to the ionization of a residue with a pK_a of 9.8 (Laws and Shore, 1978; Eftnik and Bystrom, 1986). This quenching appears to be due to the loss of 50–80% of the emission of one of two tryptophan residues of the protein with the sensitive tryptophan (Trp-314) being buried within the subunit interface of the dimeric enzyme (the insensitive tryptophan is at position 15 and appears to lie at the surface of each enzyme where it contributes very little to the total fluorescence of the protein). It has been proposed that the ionizable group responsible for this alkaline quenching is a tyrosine residue (Tyr-286). Specifically, the ionization of this residue has been suggested to quench the buried tryptophan residue by a fluorescence resonance energy transfer mechanism where the energy of the excited tryptophan is transferred to the acceptor-tyrosine group (see Section 1.3.2). A similar quenching of the intrinsic tryptophan fluorescence of the enzyme is elicited by the binding of the oxidized form of the coenzyme, NAD^+, or upon the formation of ternary enzyme–NAD^+–substrate complexes. Thus, it was proposed that the tyrosine residue at position 286 became ionized, at neutral pH, in response to NAD^+ binding. Interestingly, a lesser degree of quenching is observed upon the binding of the reduced form of the coenzyme, NADH, and this quenching can be attributed to an energy transfer process between the enzyme–tryptophan residues (emission max. = 340 nm) and the dihydronicotinamide moiety of NADH (absorption max. = 350 nm). Thus, these fluorescence studies indicate clear differences in the conformational states of the enzyme–NAD^+ and enzyme–NADH species. Overall, these types of fluorescence approaches have provided a powerful tool, in combination with crystallographic information, for obtaining a molecular understanding of the mechanisms underlying coenzyme/substrate interactions with this enzyme (Laws and Shore, 1978; Eftnik and Bystrom, 1986). It seems equally feasible that detailed molecular information regarding G protein activation and deactivation will be obtained through the application of fluorescence approaches, in conjunction with the crystal structures for these proteins which should be available in the not-too-distant future.

1.3.2. Use of Extrinsic Protein Fluorescence as a Monitor for Protein–Ligand Interactions

An alternative approach to the use of intrinsic protein fluorescence to study protein–ligand, or protein–protein, interactions is to attach an environmentally sensitive fluorophore to a protein for use as a reporter group. As alluded to above, an essential requirement of this approach is that the labeling of the protein not be deleterious to its function. The use of extrinsic reporter groups can offer some important advantages over intrinsic protein fluorescence since the quantum yields of the extrinsic probes are typically greater than that of tryptophan and their emission properties are sometimes more environmentally sensitive. Moreover, the use of carefully chosen reporter groups should enable the monitoring of key ligand-induced (or protein-induced) activities within multicomponent pathways, i.e., in those instances where the number of components makes an interpretation of the intrinsic protein fluorescence difficult.

Under normal conditions the most reactive amino acids in a protein are the cysteine and lysine residues. There are a number of examples where the attachment of a fluorescent probe to one or the other of these side chains has provided important structure–function information (Galley, 1976; Timasheff, 1970; Hiratskua and Uchida, 1980; Forgac, 1980; Gupte and Lane, 1982). Some particularly appropriate examples have involved the labeling of the Na^+/K^+-ATPase with an environmentally sensitive cysteine reagent to study the interactions of different ligands with this enzyme (Forgac, 1980; Gupte and Lane, 1982). Such studies have verified the existence of distinct conformational changes for various ligand–enzyme species on the catalytic pathway. Similar approaches, using the accessible cysteine residues on the β

subunit of transducin, should be useful for examining the individual conformational states of the G protein which occur during the receptor-stimulated activation–deactivation cycle.

Fluorescence resonance energy transfer provides an additional approach, which involves covalently attached reporter groups, for measuring protein–ligand and protein–protein interactions. The theory and application of resonance energy transfer have been put to use in a number of protein systems, where in most cases the primary aim has been to determine the relative juxtaposition of key sites on protein molecules (Hammes, 1982; Stryer, 1978). Basically, fluorescence energy transfer refers to the transfer of excitation energy from a fluorescent (donor) molecule in the excited state to a second (acceptor) chromophore in the ground (nonexcited) state. The necessary conditions for resonance energy transfer are that the fluorescence emission of the donor fluorophore overlap the absorption spectrum of an energy acceptor. The transfer of energy can be monitored by following the quenching of the donor fluorescence, or if the acceptor molecule is fluorescent, by following an increase in the emission of the acceptor. The efficiency of energy transfer (E) is dependent on the distance (R) between the donor and acceptor molecules, as described by Eq. (1), and thus

$$R = R_0(E^{-1} - 1)^{1/6} \qquad (1)$$

provides an extremely useful approach for determining the overall juxtaposition of essential domains on proteins which can be labeled with chromophoric compounds. In Eq. (1), R_0 is the distance at which the energy transfer efficiency is equal to 50% for a given donor–acceptor pair. This distance can be determined experimentally from the spectral properties of the donor and acceptor moieties. Given the relationship between the distance separating a donor–acceptor pair, and the degree of energy transfer which can occur between these moieties, it is easy to imagine how this general approach might be useful for monitoring ligand-induced conformational changes (where the structural change would be reflected by a change in the overall orientation or juxtaposition of two sites on the protein), and, in particular, for monitoring protein–protein interactions. As will be outlined in more detail in Section 2, a key issue regarding the mechanisms of activation of G proteins concerns the state of association of the subunits comprising these heterotrimeric transducer molecules. A direct approach to the study of G protein subunit association–dissociation equilibria would be to label the α and $\beta\gamma$ subunits of a G protein with an appropriate donor and acceptor pair (i.e., a donor on one subunit and the acceptor on the other) and monitor their interactions by resonance energy transfer. The expectation would be that when the α subunit is dissociated from the $\beta\gamma$ subunit complex, the distance between the donor groups on one of the subunits and the acceptor groups on the other would be great and thus yield efficiencies of energy transfer which are low or nonexistent. On the other hand, the association of the G_α subunit with the $\beta\gamma$ complex would decrease the effective distance between donors and acceptors and therefore increase the energy transfer efficiency. The potential application of this approach in studying the activation of the retinal G protein, transducin, will be discussed below.

Another issue of interest regarding the mechanisms of receptor-coupled signaling, where energy transfer approaches would be especially useful, concerns the role of receptor–receptor interactions in growth factor action. A controversial question surrounding the EGF receptor/tyrosine kinase is whether the tyrosine kinase activity is stimulated in the receptor molecule as an outcome of receptor aggregation. One possible approach to directly examine this issue involves the use of EGF molecules which are specifically labeled with appropriate fluorescence energy transfer donor–acceptor pairs. As will be described in Section 3, the EGF molecules can be specifically labeled with lysine reagents with no deleterious effects on the ability of the growth factor to bind to the receptor or to stimulate tyrosine kinase activity. Thus, strategies which are similar to those outlined above for studying G protein-subunit association and dissociation can be employed to directly monitor EGF receptor aggregation under a variety of conditions and determine whether any relationship exists between receptor aggregation and functional activity.

2. APPLICATIONS OF FLUORESCENCE TO THE STUDY OF RECEPTOR–G PROTEIN COUPLING

In Section 1.1 we outlined how a number of hormone–receptor systems regulate cellular activity through the interactions of three distinct components, these being the receptor protein itself, a GTP-binding protein which functions as a signal transducer, and an effector protein. A general scheme has emerged to explain these types of signaling pathways with the pivotal event being the receptor-stimulated binding of GTP to the α subunit of the G protein. It is this event which primes the G protein for going on to regulate the activity of a biological effector.

In fulfilling their regulatory role, the α subunits of G proteins are thought to cycle between two conformational states, an activated state induced by the binding of GTP and an inactive state which represents the GDP-bound form of this subunit. Although a great deal has been learned concerning the roles of G proteins in signal transduction, a number of important gaps remain in the general model of G protein activation and deactivation. For example, the role of G protein-subunit dissociation has not been established. It is commonly suggested that the receptor-stimulated binding of GTP to the α subunit of the G protein results in the dissociation of the αGTP species from the receptor and the $\beta\gamma$ subunit complex. However, the rates and the extent of this dissociation of the G protein subunits, during the receptor-stimulated activation–deactivation cycle in a lipid milieu, have not been directly determined. Other issues of interest include the nature of the conformational changes that the α subunit undergoes during the activation–deactivation cycle, and the factors that modulate these structural changes. The complex nature of G protein activity necessitates the development of new strategies for examining the structure–function relationships within this class of proteins.

We have outlined in Section 1 how fluorescence spectroscopy provides a powerful tool for probing structure–function studies of ligand–protein and protein–protein interactions. In the next two sections we will describe how fluorescence approaches have proven useful for studying different conformational states of G proteins as well as having provided a direct monitor for receptor–G protein coupling.

2.1. Use of Intrinsic Protein Fluorescence to Monitor the Active and Inactive Conformations of a G Protein

A variety of data point to the fact that GTP (or nonhydrolyzable GTP analogues) and GDP elicit distinct conformational states within the α subunits of G proteins. One obvious indication for this is the fact that the α subunits of G_s or transducin, which contain tightly bound molecules of GppNHp, or GTPγS, can mediate the regulation of effector activities (adenylate cyclase or the cyclic GMP PDE) while the corresponding αGDP complexes are not capable of stimulating these activities. In addition, the rate of tryptic digestion of the α subunit of transducin (α_T) is markedly changed when Gpp(NH)p is bound compared to GDP (Fung and Nash, 1983), the number of reactive sulfhydryls on α_T is reduced from three when GDP is bound to one hyperreactive residue when Gpp(NH)p is bound (Ho and Fung, 1984), and cholera toxin appears to catalyze the ADP ribosylation of the α_TGpp(NH)p complex but not the α_TGDP form (Abood et al., 1982). Thus, given the fact that each of the α subunits which have been sequenced to date contains two tryptophan residues, the intrinsic fluorescence of these proteins could provide a direct readout of the conformational changes that occur during the activation–deactivation cycle.

This approach was first utilized by Higashijima et al. (1987a) to study the intrinsic fluorescence of the purified α subunit of G_o (α_o) in detergent solution. In this study they showed that the GTPγS-bound form of α_o had an intrinsic fluorescence emission that was enhanced 15% in the presence of 1

μM GTPγS which increased to 60% when 10 mM MgSO$_4$ was added to the incubation medium. The kinetics of the fluorescence enhancement were identical to those for [^{35}S]-GTPγS binding to α_o suggesting that the increased fluorescence emission was a direct reflection of GTPγS binding. Since the binding of [^{35}S]-GTPγS to the soluble α_o was only minimally stimulated by MgSO$_4$ (1.2-fold), it appears that the bulk of the fluorescence increase upon Mg^{2+} binding to α_oGTPγS is due to a Mg^{2+}-dependent conformational change in α_oGTPγS. Thus, these fluorescence measurements apparently distinguished two different conformational changes that are involved in the activation of α_o; one which is induced by the binding of GTPγS (as reflected by a 15% increase in the tryptophan fluorescence) and the other which is dependent on Mg^{2+} binding (resulting in an additional 45% enhancement).

In further studies of the intrinsic fluorescence of α_o, Higashijima *et al.* (1987b) determined the kinetics of the fluorescence changes upon treatment with GTP or NaF/AlCl$_3$.* The results of these experiments indicated that changes in the tryptophan fluorescence of α_o could provide a direct measurement of GTP binding (i.e., as reflected by an enhancement of the intrinsic fluorescence), GTPase (as reflected by the decay of the GTP-induced fluorescence), and the activation of α_o by NaF/AlCl$_3$. In each case the maximum fluorescence change elicited was dependent on MgSO$_4$ being present. The reversibility of the fluorescence enhancement under conditions expected to reverse the activating conformation change (e.g., dilution of AlF$_4^-$ to nonactivating levels) provided further evidence that the changes in fluorescence were a direct monitor of functionally relevant structural changes. A primary advantage of the fluorescence measurement is that the kinetics of both GTP binding (the natural ligand) and GTPase activity can be measured in a single experiment which matched results from both [^{35}S]-GTPγS binding and [γ-^{32}P]-GTPase assays. Thus, it was concluded by these workers that the intrinsic protein fluorescence could serve as a direct monitor of the activation–deactivation cycle of the detergent-soluble, isolated α_o subunit. The primary disadvantage of the above system was that only the intrinsic GTP-binding GTPase cycle of the pure α_o protein, in detergent solution, was examined, i.e., the receptor- (and $\beta\gamma$-) catalyzed activation–deactivation cycle of the G protein, in a membrane environment, was not studied.

The bovine retinal phototransduction system has provided an excellent model for studying receptor–G protein–effector enzyme coupling because each of the individual components of this system can be readily purified in milligram quantities from isolated retina. Thus, the rod cell photoreceptor,

---

* It has been well documented that the combination of NaF and AlCl$_3$ elicits an activation of G proteins. This activation is felt to result from the formation of a GDP (AlF$_4^-$) complex on the α subunits (Sternweis and Gilman, 1982).

rhodopsin, can be purified from solubilized rod outer segment membranes and functionally reconstituted into phospholipid vesicles while other protein components (the G protein, transducin, and the effector enzyme, cGMP PDE) are readily soluble and can be post-added to the reconstituted rhodopsin vesicles. The use of these proteins in well-defined, reconstituted systems has already provided a good deal of insight into the basic mechanisms responsible for light-activated cGMP hydrolysis. Given the structural homology of the G proteins that couple to receptors of the class typified by rhodopsin or the β receptor, it is likely that information attained in the retinal system will be directly applicable to the other G protein-coupled signaling systems that are not as amenable to biochemical and biophysical analysis.

In an effort to utilize the advantages of the retinal phototransduction system to gain a better understanding of G protein function, we have examined whether the intrinsic tryptophan fluorescence of the α subunit of transducin (α_T) could serve as a readout for ligand-induced conformational changes in this G protein. One major advantage of this system is that transducin is eluted from the rod outer segment membranes by the addition of guanine nucleotide, so that both the GDP-bound and GTPγS-bound forms can be purified; thus, it is possible to directly compare the intrinsic fluorescence excitation and emission spectra of equal amounts of these two forms of the α_T subunit. The results of this comparison (Fig. 1) indicated that the

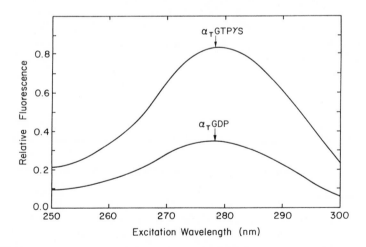

FIGURE 1. Intrinsic tryptophan fluorescence of α_TGDP and α_TGTPγS. α_TGDP (60 pmoles) and α_TGTPγS (50 pmoles) were diluted to 150 μl in HMD buffer (10 mM HEPES, pH 7.5, 5 mM MgCl$_2$, 1 mM DTT) and the fluorescence excitation spectra were scanned while monitoring emission at 330 nm. The spectra were normalized for the differences in added protein. Maximum excitation occurred at 278 nm for each form of α_T. Data from Phillips and Cerione (*J. Biol. Chem.* **263**:15498–15505).

α_TGTPγS subunit had about a twofold higher fluorescence emission (2.3 $\pm$ 0.3 S.E., $n = 10$) when comparing the excitation or emission spectra (excitation maximum 278 nm, emission maximum 334 nm) (see also Phillips and Cerione, 1988). There were no obvious differences in the fluorescence spectra of either α_T species when phosphatidylcholine vesicles were included in the incubation. Thus, the soluble, GTPγS-bound α_T subunit has a significantly greater fluorescence emission compared to the α_TGDP complex, which is similar to the results discussed above concerning the relative intrinsic fluorescence of the detergent-solubilized α_oGDP and α_oGTPγS species (Higashijima *et al.*, 1987a). Overall these fluorescence results are consistent with a number of lines of evidence (see above) indicating that the conformation of the α_TGTPγS species is significantly different from that of the α_TGDP complex.

Since the results shown in Fig. 1 suggested that the intrinsic fluorescence of the α_T subunit may serve as a sensitive monitor of the activation–deactivation cycle of the retinal G protein, we attempted to directly monitor the exchange of tightly bound GDP for GTPγS on α_T by following its fluorescence emission at 335 nm (excitation 280 nm). Thus far we have found that the fluorescence of the α_TGDP subunit, either alone or with stoichiometric amounts of $\beta\gamma_T$, does not change when incubated with GTPγS (1–100 μM) and varying amounts of MgCl$_2$ (0–100 mM) for up to 4 hr at room temperature. These results then indicated that in the absence of the photoreceptor, rhodopsin, the GDP–GTP (or GDP–GTPγS) exchange on α_T is extremely slow. This is consistent with earlier observations which indicated that the GTPase activity of transducin or isolated α_T in the absence of rhodopsin is at least 10- to 100-fold lower than the rhodopsin-stimulated GTPase activity (Cerione *et al.*, 1985). Similarly, [^{35}S]-GTPγS binding to the free α_T subunit is less than 10% complete (i.e., relative to the binding measured in the presence of light-activated rhodopsin) even after 2 hr at room temperature. It should be noted that these results are also in direct contrast to those obtained in fluorescence studies of the solubilized α_o subunit. Specifically, in the case of α_o, the binding of GTPγS is relatively rapid even in the absence of an appropriate receptor protein. Thus, there are clear differences in the intrinsic abilities of the α_o and α_T subunits to undergo guanine nucleotide exchange.

Since simple incubation with GTPγS was not sufficient to produce the active conformation of α_T, the α_TGDP subunit was treated with NaF (plus AlCl$_3$). This treatment has been suggested to produce similar conformational changes as induced by GTP binding in other G proteins (i.e., G$_s$, G$_i$, G$_o$), and in the case of G$_s$ to elicit the functional activation of adenylate cyclase (Gilman, 1987). It has been suggested that the active species is AlF$_4^-$ which binds to the αGDP subunit at the γ-phosphate binding domain, thus mimicking GTP- (or GppNHp-, GTPγS-) induced conformational changes in this sub-

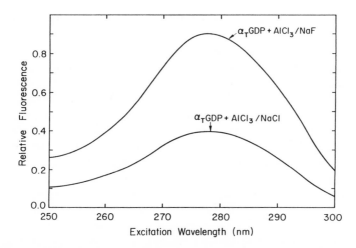

FIGURE 2. Effect of fluoride on α_TGDP and α_TGTPγS fluorescence. α_TGDP (60 pmoles) was diluted to 150 μl with HMD buffer which also contained 0.1 mM EDTA, 20 μM AlCl$_3$, and either 5 mM NaCl or 5 mM NaF. The samples were incubated at 23°C for 10 min. The excitation spectrum of each sample was then scanned and the fluorescence monitored at 335 nm. Data from Phillips and Cerione (*J. Biol. Chem.* **263**:15498–15505).

unit (Bigay *et al.*, 1985). The results shown in Fig. 2 suggest that this is in fact the case. Specifically, the addition of 5 mM NaF/10 μM AlCl$_3$ to α_TGDP results in an increase in the fluorescence emission as compared to a paired control (i.e., 5 mM NaCl/10 μM AlCl$_3$). The enhancement in this experiment was about twofold (2.2 ± 0.3 S.E., $n = 9$) which is identical to the difference in the fluorescence of the α_TGDP versus α_TGTPγS forms. The effect of the AlF_4^--induced change in fluorescence occurs within 1 min at room temperature. Similar experiments done with equal amounts of α_TGTPγS resulted in no change in the intrinsic fluorescence ($\leq$10%) indicating that the effects of AlF_4^- and GTPγS on the fluorescence of the α_T subunit were not additive. This provides further support for the suggestion that the AlF_4^- and active guanine nucleotides induce the same type of conformational change in α_T.

In an effort to compare the conformations of the α_TGDP, α_TGTPγS, and AlF_4^--treated α_TGDP subunits, the ability of a collisional quenching agent (KI) to reduce the fluorescence emission was determined. The relationship between the fluorescence of a single, homogeneous population of emitting chromophores and increasing concentrations of a quenching agent is linear and described by the Stern–Volmer equation,

$$F_0/F = 1 + K_{sv}[Q]$$

where F_0 is the fluorescence in the absence of quenching agents, F is the fluorescence at a given concentration of quenching agents, K_{sv} is the Stern–Volmer quenching constant, and [Q] is the concentration of the quenching agent. Since the effectiveness of a collisional quenching agent to reduce the fluorescence emission is dependent on a physical interaction between the excited state of the chromophore and the quencher, the accessibility of the emitting residue to solvent will be an important determinant of K_{sv}.

The results presented in Fig. 3 show that the α_TGDP species is quenched to a much greater degree compared to the α_TGTPγS complex. Also apparent is that the profiles are not simple, linear functions which suggests that either the two tryptophans of α_T contribute unequally to the overall fluorescence emission or there is heterogeneity within the protein sample (e.g., denatured α_T or nucleotide-free α_T). The important point, however, is that the AlF$_4^-$-treated α_TGDP complex is not appreciably quenched, i.e., similar to the α_TGTPγS species. In fact, the AlF$_4^-$-treated α_TGDP complex appears to be less quenched than the α_TGTPγS species suggesting that there are subtle differences in the conformations of these two α_T species. Interestingly, the AlF$_4^-$-treated α_TGDP complex will stimulate the activity of the cyclic GMP PDE; however, the α_TGDP(AlF$_4^-$) species is at best only 40–50% as effective as the α_TGTPγS species in promoting PDE activity. These results, then, provide further support for the suggestion that the α_TGDP(AlF$_4^-$) and α_TGTPγS species have different conformations. Overall, the results from the collisional quenching studies imply that the conformational change which is induced by GTPγS (or GTP), and which is somewhat mimicked by AlF$_4^-$,

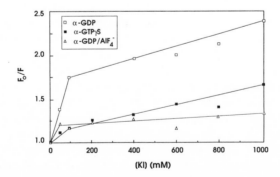

FIGURE 3. Stern–Volmer plot of KI quenching of α_TGDP and α_TGTPγS tryptophan fluorescence. α_TGDP (60 pmoles) and α_TGTPγS (75 pmoles) were diluted to 100 μl with 1.5× HMD (plus or minus 4.5 mM NaF/30 μM AlCl$_3$ in the case of α_TGDP). Fifty microliters of a 3 M salt solution varying in the proportion of KCl to KI was added to each of the α_T samples, yielding a series of diluted samples with increasing amounts of KI at a constant ionic strength. The samples were incubated at 23°C for 2 min and the maximum fluorescence was determined (excitation = 280 nm, emission = 335 nm). F_0 is the maximum fluorescence at zero KI and F is the maximum fluorescence derived at each [KI].

buries at least one of the two tryptophan residues of the α_T subunit such that it becomes shielded from the solvent KI. This, in turn, suggests that at least one of the regions of the protein that contains the tryptophan residues must be conformationally mobile.

Since Mg^{2+} has been suggested to be a necessary cofactor in the activation of G proteins (Gilman, 1987), we examined the effects of this agent on the intrinsic fluorescence of the α_TGDP subunit (Phillips and Cerione, 1988). The results of these experiments show that the AlF_4^--dependent increase in fluorescence emission is totally dependent on $MgCl_2$ (the apparent $K_d \approx 4$ mM). If the proposed mechanism of AlF_4^- activation is correct (i.e., by mimicking the binding of the γ-phosphate of GTP), the results suggest that Mg^{2+} binding is closely coupled to the binding of active guanine nucleotides, with this Mg^{2+} interaction then being essential for eliciting the activating conformational change in the α_T subunit. One possibility is that occupation of a regulatory site by Mg^{2+} induces a conformation that makes the AlF_4^- binding site (or the γ-phosphate site for active guanine nucleotides) available on α_T. The rapid kinetics of the conformational change (i.e., <1 min) at 10 mM $MgCl_2$ suggest that the site is readily available. The effect of $MgCl_2$ on the tryptophan fluorescence of the α_TGDP species in the presence of 2 mM EDTA is ~30% greater than when net mM concentrations of $MgCl_2$ are present. The Mg^{2+}-induced quenching of the fluorescence of the free α_TGDP species occurs over the same range as the Mg^{2+}-dependent increase in the fluorescence of the α_TGDP (AlF_4^-) species. Whereas we never see an effect of AlF_4^- on the fluorescence of the α_TGTPγS subunit, we do see a similar Mg^{2+}-dependent quenching when α_TGTPγS is titrated with increasing concentrations of $MgCl_2$. Thus, these results indicate that there is a $MgCl_2$ binding site that modulates the conformations of both the GDP and GTPγS-bound forms of the α_T subunit. While Mg^{2+} binding to the α_o subunit also has a significant effect on the conformation of that subunit, these metal-induced conformational changes clearly differ from those observed in α_T. Specifically, Mg^{2+} elicits an additional enhancement of the intrinsic tryptophan fluorescence of the α_oGTPγS complex and these Mg^{2+}-induced changes in fluorescence appear to be specific for the active form of α_o. Thus, while the binding of Mg^{2+} to the α subunits of G proteins may be a general regulatory feature of these transducers, the exact nature of the metal-induced conformational change may vary among the different α subunits.

2.2. Use of Intrinsic Protein Fluorescence to Monitor the Receptor-Stimulated Activation and Deactivation of G Proteins

As outlined in Section 1.1, the use of the purified components of the retinal phototransduction system in well-defined reconstituted systems has already provided a great deal of insight into the basic mechanisms responsible for light-activated cyclic GMP hydrolysis. Nonetheless, a number of questions remain regarding the detailed sequence of events operating in the receptor-stimulated activation–deactivation cycle of the G protein. Among these questions are the exact points in the cycle where guanine nucleotide exchange and G protein subunit dissociation occur and the steps in the cycle which are specifically influenced by the photoreceptor, rhodopsin, by the β and γ subunits of the G protein, and by $MgCl_2$. In order to develop real time assays for these receptor/G protein-coupled events we have examined the possibility of using fluorescence spectroscopy to directly monitor the reconstituted interactions between pure rhodopsin and transducin in phospholipid vesicle systems. An obvious complication in using intrinsic tryptophan fluorescence to monitor these events is that both rhodopsin and $\beta\gamma$ (which is necessary for coupling α_TGDP to the receptor) contain tryptophan residues. Thus, the measurements of changes in the α_T emission will always be against a background of fluorescence that must be controlled for and eventually subtracted to yield the net contribution of α_T. We use an SLM 8000 fluorometer and data are collected with an IBM PC using the SLM software package that controls data collection, storage, and manipulation. This allows for relatively easy corrections of fluorescence spectra by scanning and storing the background fluorescence (i.e., $\beta\gamma_T$ and rhodopsin vesicles) which subsequently can be subtracted from the total fluorescence (i.e., after the addition of α_T). In this way the fluorescence of α_T can be derived providing that the fluorescence of $\beta\gamma_T$ and rhodopsin remain relatively constant.

The results in Fig. 4A show the GTPγS-induced effects on the net fluorescence spectra of α_T, derived from the total fluorescence of a reconstituted system containing rhodopsin vesicles, $\beta\gamma_T$, and α_TGDP. The results indicate that the excitation spectrum of the α_T subunit increases about 1.7-fold when the cuvette was exposed to room light for 30 min. When a similar incubation was done with GDP instead of GTPγS (Fig. 4B), the derived α_T spectra were identical whether or not the rhodopsin vesicles were exposed to light. Based on the results that the pure α_TGTPγS has about a 2-fold greater intrinsic fluorescence than α_TGDP (see above), the enhancement shown in panel A most likely reflects the rhodopsin-catalyzed binding of GTPγS to α_T. In further experiments it was shown that the light- and GTPγS-dependent increase in the fluorescence emission was totally dependent on both added rhodopsin and $\beta\gamma_T$. It should also be noted that the incubation of reconstituted rhodopsin and $\beta\gamma_T$, alone (i.e., without α_TGDP), with GTPγS showed no time-dependent changes in fluorescence. Taken together, these data indicate that under conditions which support the activation of the α_TGDP subunit (i.e., the conversion of α_TGDP to α_TGTPγS), a specific increase in the total intrinsic fluorescence is observed which appears to be a direct outcome

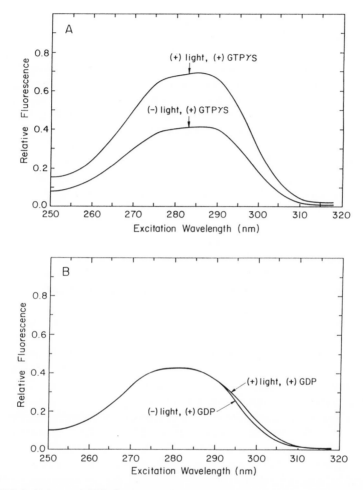

FIGURE 4. Light- and GTPγS-stimulated enhancement of the intrinsic fluorescence of α_TGDP. (A) The corrected fluorescence excitation spectra for α_TGDP ($\approx$400 nM), obtained in the light and in the dark, following the addition of GTPγS to rhodopsin-containing vesicles (rhodopsin $\approx$ 35 nM), α_TGDP, and $\beta\gamma_T$ ($\approx$400 nM) in HMD buffer. The final concentration of GTPγS was 130 μM. (B) The same experiment was performed as in A except that 130 μM GDP was added rather than GTPγS. Data from Phillips and Cerione (*J. Biol. Chem.* **263**:15498–15505).

of the receptor-stimulated formation of the α_TGTPγS complex. Thus, this system provides a means of directly monitoring transducin activity in a well-defined, receptor-coupled reconstituted system.

Since all of the above fluorescence measurements were made after prolonged incubation times with the nonhydrolyzable GTP analogue, GTPγS, we wanted to determine whether the complete activation–deactivation cycle of transducin (GTP-binding GTPase) could be monitored fluorimetrically. In these experiments the total tryptophan fluorescence (excitation 280 nm, emission 335 nm) was continuously monitored, and the rhodopsin-catalyzed GDP–GTP exchange reaction was initiated by the addition of guanine nucleotide. In order to determine the fluorescence contribution of α_T from the total protein fluorescence, incubations were set up with reconstituted rhodopsin vesicles, $\beta\gamma_T$, and guanine nucleotide, and the level of intrinsic fluorescence was determined. Since the fluorescence of all incubations which lacked added α_T was relatively stable over the time period of the experiment, the component of the total fluorescence due to the α_T subunit was easily derived by subtraction of the fixed background fluorescence (i.e., due to rhodopsin and $\beta\gamma_T$) over the same time period.

Figure 5A shows the net fluorescence of the α_T subunit in the presence of light-activated rhodopsin vesicles, $\beta\gamma_T$, and GTP as a function of time. The addition of GTP results in a rapid increase in the fluorescence emission that reaches a maximum level in approximately 1 min. The peak of enhancement is then followed by a steady decrease in the emission that is complete within 5–10 min following the addition of GTP. Interestingly, the level of fluorescence emission following the quenching phase is about 30% greater than the baseline prior to GTP addition. Although the reason for this is not known, a similar effect was noted by Higashijima *et al.* (1987a) using the pure α_o subunit (i.e., not in the presence of receptor and $\beta\gamma$). This may suggest that the α subunit has a slightly different conformation after a single turnover of the activation–deactivation cycle.

The immediate enhancement in fluorescence is most likely due to GTP binding while the subsequent decay phase reflects GTP hydrolysis. This is supported by the results shown in Fig. 5B which demonstrate that if GTPγS was added instead of GTP there was a rapid fluorescence enhancement that remained stable (i.e., no quenching phase) during the time course of the experiment. This suggests that the inability of GTPγS to be hydrolyzed directly correlates to the observation of a GTPγS-induced enhancement of the intrinsic fluorescence of α_T which persists over several minutes. A number of further experiments also support this conclusion: (1) a second addition of GTP following the decay phase elicited another sequence of fluorescence enhancement and decay while no further change was observed when GTP was added to the GTPγS-treated system; (2) addition of an activating amount of AlF$_4^-$ after the decay phase produced a rapid fluorescence enhancement while no further change was observed when GTPγS was used. These data strongly support the conclusion that the change in net fluorescence attributable to the α_T subunit is a direct, real time measurement of the GTP-binding GTPase (i.e., activation–deactivation) cycle of transducin.

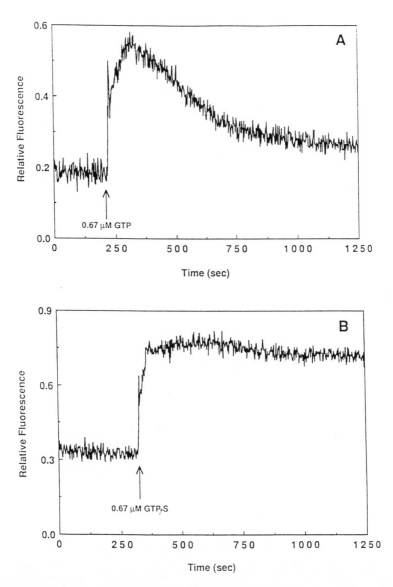

FIGURE 5. Kinetics of the light-stimulated fluorescence enhancement of reconstituted α_TGDP. (A) A mixture containing rhodopsin vesicles (rhodopsin $\approx$ 10 nM), $\beta\gamma_T$ ($\approx$300 nM), and α_TGDP ($\approx$400 nM) in HMD buffer was incubated in room light for 10 min at 23°C, and the baseline fluorescence measured (excitation $\approx$ 280 nm, emission $\approx$ 335 nm). At the time indicated by the arrow, GTP was added to a final concentration of 625 nM and rapidly mixed while continuing to monitor the emission. The trace shown in A is the time-dependent emission of the added α_T that was derived by subtracting the fluorescence emission contributed by rhodopsin and $\beta\gamma_T$ over the same time period. (B) Experimental conditions were identical to those in A except that an equivalent amount of GTPγS was added instead of GTP. Data from Phillips and Cerione (1988).

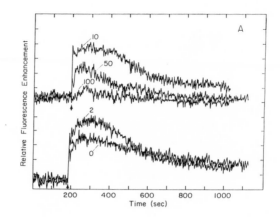

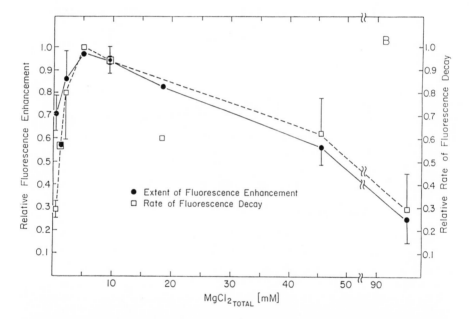

FIGURE 6. Effect of MgCl$_2$ on the light- and GTP-stimulated changes in the intrinsic fluorescence of α_TGDP. (A) A mixture of rhodopsin-containing phospholipid vesicles (rhodopsin $\approx$ 10 nM), $\beta\gamma_T$ ($\approx$300 nM), and α_TGDP ($\approx$400 nM) in 10 mM HEPES (pH 7.5), 1 mM DTT, 2 mM EDTA, and containing the indicated amounts of MgCl$_2$ were incubated in room light for 10 min at 23°C. The fluorescence emission was monitored at 335 nm (excitation 280 nm) and GTP was added to a final concentration of 625 nM as shown by the arrow during continuous monitoring. The traces represent the emission of α_T corrected for the fluorescence contributed by rhodopsin and $\beta\gamma_T$. (B) Plots of the final extent of fluorescence enhancement and the rate of fluorescence decay as a function of the total MgCl$_2$ present in the cuvette solutions. The extent of fluorescence enhancement was determined from the final plateau values for the different fluorescence traces shown in A. The rate of fluorescence decay was measured from the linear slope of the fluorescence quenching measured between the 400 and 600 sec time period for the traces shown in A.

As discussed above, since $MgCl_2$ is known to modulate G protein structure and function, we wished to use the fluorescence assay of the reconstituted system to determine the specific effects of Mg^{2+} on the rhodopsin-stimulated activation–deactivation cycle of transducin. Figure 6A shows the time-dependent changes in the corrected intrinsic fluorescence of $\alpha_T GDP$ in response to the addition of GTP to a light-activated, reconstituted system at various concentrations of $MgCl_2$. These data show that maximum enhancement (GTP binding) occurs within 5 mM (total) $MgCl_2$, and most of the change (70–75%) occurs under conditions where $[MgCl_2] \ll 1$ mM (i.e., total $[MgCl_2] = 0.3$ mM, $[EDTA] = 2$ mM). The traces also show that at the higher $MgCl_2$ concentrations the fluorescence enhancement was diminished. The effect of $MgCl_2$ on the relative fluorescence enhancement (GTP binding) and the rate of fluorescence decay (GTPase) were determined from the traces in Fig. 6A and a second similar experiment (not shown) and these data are plotted in Fig. 6B. One obvious result of this analysis is that there was a much greater effect of $MgCl_2$ (up to 5 mM total) on the fluorescence decay, reflective of GTPase activity (i.e., 3-fold increase), than on the level of enhancement (1.4-fold increase) due to GTP binding. Higher concentrations of $MgCl_2$ (20–100 mM) produced identical inhibitions of both fluorescence enhancement and decay. Thus, there appears to be a relatively tight $MgCl_2$ interaction with α_T which is complete in the low millimolar concentration range, and this has a significant influence on the intrinsic rate of GTP hydrolysis. As shown above (see Fig. 4), $MgCl_2$ affects the intrinsic fluorescence of $\alpha_T GDP$ (i.e., induces a quenching of the fluorescence), and promotes the AlF_4^--induced enhancement in the $\alpha_T GDP$ fluorescence, with a similar concentration dependence as that seen for the promotion in the rate of the GTP hydrolytic event. It is worth noting that added Mg^{2+} does not appear to be essential for promoting the activation (i.e., the GTP binding) of the α_T subunit (i.e., the bulk of the activating conformational change still occurs in 2 mM EDTA). This clearly seems to be in contrast to the situation for α_o where it has been proposed that the activation of this subunit requires the binding of GTP, or a nonhydrolyzable GTP analogue, together with Mg^{2+} (Higashijima et al., 1987a,b).

A second interesting feature of the data presented in Fig. 6 is that both the fluorescence enhancement and the decay phases are inhibited to a similar degree by $MgCl_2$ concentrations ranging from 10 to 100 mM. It has been shown previously that high $[MgCl_2]$ will induce the dissociation of transducin into free α_T and $\beta\gamma_T$ subunits (Deterre et al., 1984). Since it is the holotransducin form that functionally interacts with the receptor, the inability to form the holoprotein could explain the progressive loss of the fluorescence enhancement. The fact that the GTPase rate was similarly inhibited [as measured by fluorescence quenching and $^{32}P_i$ release assays (data not shown)] suggests that this Mg^{2+}-induced effect may be localized to the α_T subunit. These data seem to highlight a primary advantage in using fluorescence spectroscopy to monitor receptor-stimulated G protein activation and deactivation; specifically it enables the simultaneous monitoring of the effects of a given component (such as Mg^{2+}) on both the activation (GTP binding) and deactivation (GTP hydrolysis) events in the same experiment. This, then, enables a clear dissection of which step is being affected by a given agent in the GTPase cycle.

Thus, in the above sections, we have described how the intrinsic fluorescence of a G protein can be used to monitor structural and functional changes in response to receptor coupling. Specifically, we have shown that the changes in the intrinsic fluorescence of α_T occurred under conditions known to produce the activation and subsequent deactivation of α_T. Future studies will be aimed at a closer examination of the determinants of the important conformational change(s) that occur during the cycle of α_T, and the influence that varying levels of rhodopsin, $\beta\gamma_T$, and/or cGMP PDE may have on the different phases and kinetics of the activation–deactivation cycle of α_T.

2.3. Use of Extrinsic Fluorescent Probes to Monitor Transducin Activity

A second approach to directly monitoring transducin activity is through the use of fluorescent reporter groups covalently bound to one or both of the subunits of transducin. The primary goal of this approach is to covalently modify the subunit(s) with a fluorescent probe that does not inhibit the function of the protein. It has already been shown that two cysteine residues on $\beta\gamma_T$ can be modified with 5,5'-dithiobis-(2-nitrobenzoic acid) (DTNB) with no loss of light-stimulated GTP binding or GTPase activity of the α_T subunit (Ho and Fung, 1984). Thus, we set out to determine if the $\beta\gamma_T$ sulfhydryls could be modified with environmentally sensitive fluorescent probes that would change their fluorescent spectra when complexed with α_T, lipid, and/or receptor. If a fluorescent-labeled $\beta\gamma_T$ subunit could serve as a structure–function readout of this subunit complex, then important questions concerning the role of G protein-subunit association and dissociation could be directly approached in a receptor-coupled reconstituted system.

The purified $\beta\gamma_T$ subunit (50 μM) was reacted with 150 μM 6-acryloyl-2-dimethylaminonaphthalene (acrylodan), and the sulfhydryl modification was followed as either the increase in the 510-nm emission (excitation 390 nm) or the quenching of the intrinsic tryptophan fluorescence (excitation 280 nm, emission 335 nm) as shown in Fig. 7. The increase in the emission of the acrylodan fluorescence follows a time course similar to that previously

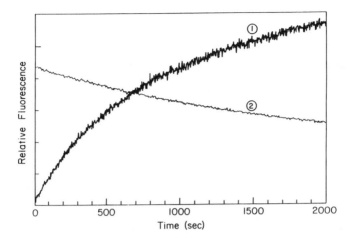

FIGURE 7. Modification of $\beta\gamma_T$ with acrylodan. Trace 1: purified $\beta\gamma_T$ (12 μM) in 20 mM HEPES (pH 7.5), 5 mM MgCl$_2$, 0.15 M NaCl, and 25% glycerol was mixed with acrylodan to yield a final concentration of 150 μM. The acrylodan was diluted from a 10 mM stock solution made up in dimethylformamide. The fluorescence emission at 510 nm (excitation 390 nm) of the acrylodan label was continuously monitored at one determination per second with the fluorometer operating in the ratio (i.e., A/B) mode. Trace 2: the experiment was identical to that in trace 1 except that the tryptophan fluorescence emission at 335 nm (excitation 280 nm) was monitored over the same period, and the $\beta\gamma_T$ concentration was 20 μM.

shown for DTNB modification of $\beta\gamma_T$ (Ho and Fung, 1984). When the reaction was stopped by the addition of 1 mM DTT, and the reaction products separated on a desalting column (Biogel P6DG), the stoichiometry of incorporation was generally 2–3 moles acrylodan per mole $\beta\gamma_T$. The observation that the quantum yield of the label increased as a function of the time suggested that the sites being modified were somewhat hydrophobic and shielded from the aqueous solvent. The fact that denatured $\beta\gamma_T$ has six cysteines which react within 10 min of DTNB addition (Ho and Fung, 1984) suggested that the slow time course of the native $\beta\gamma_T$ modification was due to the sulfhydryls being somewhat inaccessible to the surrounding medium. Interestingly, the increase in acrylodan fluorescence emission correlated with a somewhat slower but significant quenching of the $\beta\gamma_T$ tryptophan fluorescence (Fig. 8). One possibility for this quenching is fluorescence energy transfer from the emitting tryptophans (donor) to the absorbing acrylodan residues (acceptor). Support for this comes from data (unpublished observations) showing a time-dependent increase in 510-nm emission when the tryptophans are excited at 280 nm. Taken together, these findings indicated that the $\beta\gamma_T$ complex has been specifically modified with an environ-

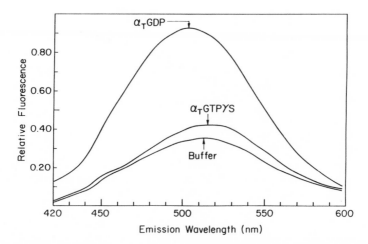

FIGURE 8. Fluorescence spectra of acrylodan-labeled $\beta\gamma_T$. The purified $\beta\gamma_T$ complex was reacted with a 12-fold molar excess of acrylodan for 2 hr at 23°C in a buffer of 20 mM HEPES (pH 7.5), 5 mM $MgCl_2$, 0.15 M NaCl, and 25% glycerol. At 2 hr the unreacted acrylodan was quenched by addition of 2 mM DTT, and the protein fraction was separated from the low-molecular-weight adducts by Biogel P6DG desalting gel chromatography. The protein peak was pooled and the stoichiometry of incorporation was determined to be 2.4 moles per mole $\beta\gamma_T$. The labeled $\beta\gamma_T$ (15.6 pmoles) was incubated with either 86 pmoles α_TGDP, 120 pmoles α_TGTPγS, or α_T buffer for 10 min at 23°C, and the primary acrylodan fluorescence emission spectrum (excitation 390 nm) of each was determined. The traces shown have been corrected for background fluorescence which was determined for each incubation in the absence of acrylodan-$\beta\gamma_T$ and eventually subtracted from the primary spectra.

mentally sensitive fluorescent probe, and further work was directed at evaluating the labeled $\beta\gamma_T$ as a specific probe of transducin activity.

Figure 8 shows the fluorescence spectra of acrylodan-labeled $\beta\gamma_T$ in the presence of buffer, α_TGDP, or α_TGTPγS. The results of these incubations show that the emission of the acrylodan label is significantly increased (2–3 fold) in the presence of α_TGDP but not α_TGTPγS or α_T buffer. The simplest interpretation is that the association of acrylodan-$\beta\gamma_T$ with α_TGDP results in a further shielding of the fluorophore resulting in an increased quantum yield along with a blueshift in the wavelength maximum emission. It has been previously suggested that the activation of transducin (i.e., upon the binding of GTP or GTPγS) results in the dissociation of its subunits, due to a greatly reduced affinity of α_T for $\beta\gamma_T$ (Fung, 1983). This would provide an explanation for the lack of an increase in the emission of the label in the presence of α_TGTPγS and would strongly support the current model of subunit dissociation upon activation. These data also suggest that changes in the acrylodan fluorescence of the labeled $\beta\gamma_T$ may provide a specific probe for the association and dissociation of α_T and $\beta\gamma_T$. Future studies will be directed at determining the specificity of the fluorescence changes under a variety of conditions including an evaluation of the effect of lipid on the fluorescence. The goal of these studies will be to develop a reconstituted system in which the association and dissociation of the labeled $\beta\gamma_T$ can be followed directly under carefully controlled conditions. With these systems it will be possible to test a number of the unproven tenets of the current model of G protein activity such as the kinetics of subunit dissociation during the activation of α_T and the necessity of $\beta\gamma_T$ dissociation from receptor to achieve its catalytic activity. Thus, the ability to synthesize specifically labeled probes of transducin should provide new insight into the basic mechanisms of G protein activity.

3. FUTURE DIRECTIONS—USE OF FLUORESCENCE SPECTROSCOPY TO STUDY THE RELATIONSHIP BETWEEN RECEPTOR–RECEPTOR INTERACTIONS AND RECEPTOR ACTIVATION

A central issue regarding growth factor-mediated signal transduction concerns how growth factor binding to the extracellular domain of the receptor can effect functional changes at the cytoplasmic tyrosine kinase domain. Two different mechanisms have been proposed to account for this transmembranal signaling event. One involves a conformational change in the receptor which is elicited by growth factor binding to the extracellular domain but which is propagated across the membrane to the tyrosine kinase domain, i.e., an intramolecular signaling mechanism. This type of a ligand-induced conformational change is an essential component of other receptor-coupled signaling systems; for example, hormone binding to the βAR triggers a conformational change within this receptor which promotes its interaction with the G_s protein. Similarly, light absorption by rhodopsin stimulates its interaction with the retinal G protein, transducin. However, in both of these cases the primary amino acid sequences suggest that the receptor proteins are comprised of a number of membrane, helical spanning regions (i.e., 7), clearly differing from the EGF receptor/tyrosine kinase (or other receptor/tyrosine kinases, i.e., insulin receptor, platelet-derived growth factor receptor) which is comprised of a single, narrow transmembranal stretch of amino acids (i.e., 23) (Ullrich *et al.*, 1984). These structural considerations have in fact prompted the question of whether transmembranal conformational changes can be transmitted across the narrow membrane spanning region connecting the EGF binding and tyrosine kinase domains. Thus, a second mechanism for growth factor regulation of tyrosine kinase activity has been suggested where EGF binding to the extracellular domain triggers receptor–receptor interactions that in turn result in a stimulation of the catalytic activity of the tyrosine kinase domain.

A good deal of evidence has been gathered which supports both of these mechanisms. For example, studies using antibodies prepared against the EGF receptor, and experiments monitoring the mobility of the receptors on intact cells through fluorescence photobleaching recovery approaches (Schlessinger et al., 1978; Schechter et al., 1979; Zidovetzki et al., 1981; Schreiber et al., 1981, 1983; Yarden and Schlessinger, 1987a,b), have led Schlessinger and colleagues to suggest that receptor aggregation (clustering) is necessary for the stimulation of DNA synthesis and mitogenesis. However, hydrodynamic studies performed on purified EGF receptors in detergent solution by Das and colleagues (Biswas et al., 1985; Basu et al., 1986), as well as in our own laboratory (Koland and Cerione, 1988), seem to indicate that the monomeric form of the receptor functions as an active tyrosine kinase. Similarly, experiments performed with the purified insulin receptor, which in essence represents a dimeric version of the EGF receptor (and is roughly twice its molecular weight), also suggest that receptor–receptor interactions are not required for the induction of tyrosine kinase activity (Shia et al., 1983; Petruzelli et al., 1984), although interactions between the α–β dimers that comprise the α_2–β_2 receptor molecule may play some regulatory role in this activity (Boni-Schnetzler et al., 1986, 1987). Overall, a major difficulty with the results obtained using purified preparations of the growth factor receptor/tyrosine kinases is that these studies have been performed in the presence of detergent which may compromise the physiological relevance of the results. Specifically, the detergents could interfere with protein–protein (receptor–receptor) interactions that might normally occur within a detergent-free lipid environment.

One potential direct approach toward determining whether receptor–receptor interactions can occur in a variety of environments (cell membranes, reconstituted phospholipid vesicle systems, detergent solution) is through the use of fluorescence resonance energy transfer techniques using growth factor molecules which have been labeled with fluorescent reporter groups. As outlined in Section 1.3.2, the efficiency of energy transfer between a donor fluorophore and an acceptor moiety is dependent on the distance separating these molecules. Thus, the specific experimental strategy would involve incubating the EGF receptors with sufficient EGF molecules labeled with an energy transfer acceptor moiety to titrate 70–80% of the total receptor sites. A sufficient amount of EGF molecules labeled with an energy transfer donor would be included in these incubations to titrate the remaining receptor sites. The expectation would be that if the labeled EGF molecules do not induce the formation of receptor aggregates, then the distance between the donor groups on any one EGF molecule and the acceptor groups on another molecule would be great and there would be little or no measurable energy transfer between the labeled growth factor molecules. However, if the EGF-donor molecules interact with an aggregated receptor, it would be highly likely (given the experimental conditions just described) that an EGF-acceptor molecule would be bound to the other receptor molecule(s) in the aggregate. This would be reflected by a quenching of the donor fluorophore or by an enhancement of the emission of the acceptor moiety, provided that the distance separating the labeled growth factor molecules within the aggregate are not sufficiently greater than the R_o value (i.e., $>1.5 \times R_o$) for the donor–acceptor pair. Some especially good donor–acceptor pairs for these purposes are fluorescein and eosin or fluorescein and rhodamine, since the fluorescence quantum yield of the fluorescein and the extinction coefficients for eosin and rhodamine are high and yield R_o values in the range of 50–60 Å. Thus, as long as these donor–acceptor moieties are within 75 Å of each other, some degree of energy transfer should be observed. We, in fact, have prepared the fluorescein-, eosin-, and rhodamine-labeled EGF molecules by attaching the isothiocyanate derivatives of these chromophores to EGF molecules (Carraway *et al.,* 1989). This labeling was performed by exploiting the fact that the only reactive amino side chain on the EGF molecule is the NH_2-terminus (Chatelier *et al.,* 1986). An additional advantage is that this region of the growth factor molecule does not appear to participate in the binding of the growth factor to the receptor and so it was anticipated that the modification of this domain would not interfere with growth factor function. This turns out to be the case. The EGF molecules can be labeled with these fluorescent isothiocyanates with stoichiometries of 0.9–1.4 moles fluorophore per mole EGF and these labeled molecules will bind to membrane preparations of the receptor and will stimulate the tyrosine kinase activities of these preparations to extents comparable to the stimulations elicited by native EGF. Recently, we have observed that when A431 membranes are incubated with 1 nM EGF-fluorescein isothiocyanate (FITC) and 5 nM EGF-rhodamine isothiocyanate (RITC), resonance energy transfer is observed between the labeled growth factor molecules (Fig. 9), as reflected by a quenching of the component of the total fluorescence due to the EGF-FITC molecule and the corresponding increase in the sensitized emission of the EGF-RITC. The energy transfer can be blocked by preincubating the membranes with excess native EGF, indicating that this is a receptor-specific phenomenon, apparently due to a receptor–receptor interaction which brings the two probes into sufficient proximity to undergo energy transfer. Similar results have been obtained using EGF-FITC and EGF-eosin isothiocyanate as a donor–acceptor pair; a careful characterization of these effects indicated that receptor–receptor interactions occurred to a maximal extent

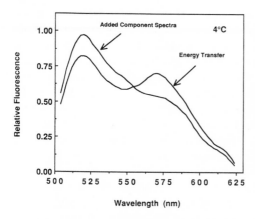

Wavelength (nm)

FIGURE 9. Resonance energy transfer between labeled EGF molecules bound to A431 membranes. The fluorescence emission spectra (excitation 490 nm) were obtained for a mixture of 5 nM FITC-EGF and 50 nM RITC-EGF in 20 mM HEPES (pH 7.4), 0.25 mg/ml (protein) A431 membranes (designated in the figure by the arrow labeled "Energy Transfer"). The individual spectra obtained for FITC-EGF, alone, in the presence of A431 membranes, and RITC-EGF, alone (in the presence of membranes) were summed and are indicated by the arrow labeled "Added Component Spectra."

when divalent metal ion activators (Mn^{2+} or Mg^{2+}) were added to the membranes together with the labeled EGF molecules (Carraway *et al.*, 1989). Thus, under conditions where the EGF receptor molecules are primed to be active as tyrosine kinases, they appear to be aggregated, at least on the surface of purified A431 cell membranes. Studies are now being directed toward comparing the degrees of energy transfer observed in membranes to those recorded in intact A431 cells and in reconstituted phospholipid vesicles containing the purified EGF receptor.

It should be noted that while the use of these labeled EGF molecules should provide a good deal of information regarding receptor–receptor interactions, ultimately it will be necessary to label the receptor molecules with appropriate donor–acceptor reporter groups in order to determine whether the growth factor itself triggers receptor–receptor interactions, under any set of conditions. This represents a tall order for the EGF receptor system, which unlike the case for the primary components of the retinal visual system, is not readily available in large (mg) quantities. However, procedures are under way in a number of laboratories to develop expression systems for the receptor that would be amenable to purifying large quantities of protein. In terms of a specific side chain candidate for modification, the EGF receptor has been shown to contain a highly reactive cysteine residue which can be modified in a matter of minutes. The outcome of this modification is an essentially complete inhibition of all tyrosine kinase activity. We find that the rates and extents of this rapid cysteine modification are not inhibited by preincubation of the EGF receptor with growth factor or divalent metals (in fact, Mn^{2+} stimulates the reactivity of this cysteine*) and are only slightly

---

* Koland, J. G., and Cerione, R. A., 1990, Mechanism of activation of EGF receptor kinase activity by divalent metal ions: A comparison of holoreceptor and isolated kinase domain properties, *Biochimica et Biophysica Acta* **1052**:489–498.

reduced by saturating levels of ATP. These results suggest that the essential cysteine residue is not present at the binding domains of these different agents; thus, the selective labeling of this residue could yield a fluorescent receptor molecule that might be used for studying growth factor or metal-induced conformational changes in the receptor and/or receptor–receptor interactions. Recently, we have shown that the receptor can be rapidly labeled with a fluorescent maleimide derivative (MIANS).* Future studies will be directed toward preparing sufficient labeled receptor for energy transfer studies like those presently under way with the labeled growth factor molecules.

4. REFERENCES

Abood, M. E., Hurley, J. B., Pappone, M.-C., Bourne, H. R., and Stryer, L., 1982, *J. Biol. Chem.* **257**:10540–10543.
Asano, T., and Ross, E. M., 1984, *Biochemistry* **23**:5467–5471.
Asano, T., Pedersen, S. E., Scott, C. W., and Ross, E. M., 1984, *Biochemistry* **23**:5460–5467.
Baehr, W., Devlin, M. J., and Applebury, M. L., 1979, *J. Biol. Chem.* **254**:11669–11677.
Basu, M., Majumdar-Sen, A., Basu, A., Murthy, U., and Das, M., 1986, *J. Biol. Chem.* **261**:12879–12882.
Benovic, J. L., Shorr, R. G. L., Caron, M. G., and Lefkowitz, R. J., 1984, *Biochemistry* **23**:4510–4518.
Berridge, M. J., and Irvine, R. F., 1984, *Nature* **312**:315–321.
Bigay, J., Deterre, P., Pfister, C., and Chabee, M., 1985, *FEBS Lett.* **191**:181–185.
Biswas, R., Basu, M., Majumdar-Sen, A., and Das, M., 1985, *Biochemistry* **24**:3795–3802.
Bokoch, G. M., and Gilman, A. G., 1984, *Cell* **39**:301–308.
Boni-Schnetzler, M., Rubin, J. B., and Pilch, P. F., 1986, *J. Biol. Chem.* **261**:15281–15287.
Boni-Schnetzler, M., Scott, W., Waugh, S. M., DiBella, E., and Pilch, P., 1987, *J. Biol. Chem.* **262**:8395–8401.
Brand, L., and Witholt, B., 1967, *Methods Enzymol.* **11**:776–856.
Branden, C.-I., Jornvall, H., Eklund, H., and Furugren, B., 1975, in *The Enzymes* (P. D. Boyer, ed.), Academic Press, New York, pp. 103–190.
Brandt, D. R., Asano, T., Pedersen, S. E., and Ross, E. M., 1983, *Biochemistry* **22**:4357–4362.
Burch, R. M., Luini, A., and Axelrod, J., 1986, *Proc. Natl. Acad. Sci. USA* **83**:7201–7205.
Carraway, K. L., Koland, J. G., and Cerione, R. A., 1989, *J. Biol. Chem.* **264**:8699–8707.
Cerione, R. A., Codina, J., Benovic, J. L., Lefkowitz, R. J., Birnbaumer, L., and Caron, M. G., 1984a, *Biochemistry* **23**:4519–4525.
Cerione, R. A., Sibley, D. R., Codina, J., Benovic, J. L., Winslow, J., Neer, E. J., Birnbaumer, L., Caron, M. G., and Lefkowitz, R. J., 1984b, *J. Biol. Chem.* **259**:9979–9982.
Cerione, R. A., Staniszewski, C., Benovic, J. L., Lefkowitz, R. J., Caron, M. G., Gierschik, P., Somers, R., Spiegel, A. M., Codina, J., and Birnbaumer, L., 1985, *J. Biol. Chem.* **260**:1493–1500.
Cerione, R. A., Benovic, J. L., Codina, J., Birnbaumer, L., Lefkowitz, R. J., and Caron, M. G., 1986, in *The Receptors* (P. M. Conn, ed.), Academic Press, New York, pp. 2–34.
Cerione, R. A., Gierschik, P., Staniszewski, C., Benovic, J. L., Codina, J., Somers, R., Birnbaumer, L., Spiegel, A. M., Lefkowitz, R. J., and Caron, M. G., 1987, *Biochemistry* **26**:1485–1491.
Cerione, R. A., Kroll, S., Rajaram, R., Unson, C., Goldsmith, P., and Spiegel, A. M., 1988, *J. Biol. Chem.* **263**:9345–9352.

Chatelier, R. C., Ashcroft, R. G., Lloyd, C. J., Nice, E. C., Whitehead, R. H., Sawyer, W. H., and Burgess, A. H., 1986, *EMBO J.* **5:**1181–1186.

Codina, J., Yatani, A., Grenet, D., Brown, A. M., and Birnbaumer, L., 1987, *Science* **236:**442–445.

Deterre, P., Bigay, J., Pfister, C., and Chabre, M., 1984, *FEBS Lett.* **178:**228–232.

Dixon, R. A. F., Kobilka, B. K., Strader, D. J., Benovic, J. L., Dohlman, H. G., Frielle, T., Bolanowski, M. A., Bennett, C. D., Rands, E., Diehl, R. E., Mumford, R. A., Slater, E. E., Sigal, I. S., Caron, M. G., Lefkowitz, R. J., and Strader, C. D., 1986, *Nature* **32:**75–79.

Eftnik, M. R., and Bystrom, K., 1986, *Biochemistry* **25:**6624–6630.

Forgac, M. D., 1980, *J. Biol. Chem.* **255:**1547–1553.

Fung, B. K.-K., 1983, *J. Biol. Chem.* **258:**10495–10502.

Fung, B. K.-K., and Nash, C. R., 1983, *J. Biol. Chem.* **258:**10503–10510.

Galley, W. C., 1976, in *Biochemical Fluorescence: Concepts,* Vol. 2 (R. F. Chen and H. Edehoch, eds.), Dekker, New York, pp. 409–436.

Gawler, D., and Houselay, M. D., 1987, *FEBS Lett.* **216:**94–98.

Gilman, A. G., 1987, *Annu. Rev. Biochem.* **56:**615–650.

Gomperts, B. D., 1983, *Nature* **306:**64–66.

Gupte, S. S., and Lane, L. K., 1982, *J. Biol. Chem.* **257:**5005–5012.

Hammes, G. G., 1982, *Enzyme Catalysis and Regulation,* Academic Press, New York, pp. 20–26.

Hargrave, P. A., McDowell, J. H., Curtis, D. R., Wang, J. K., Juszczak, E., Fong, L.-L., Rao, J. K. M., and Argos, P., 1983, *Biophys. Struct. Mech.* **9:**235–244.

Hescheler, J., Rosenthal, W., Trautwein, W., and Schultz, G., 1987, *Nature* **325:**445–447.

Heyworth, C. M., and Houselay, M. D., 1983, *Biochem. J.* **214:**547–552.

Heyworth, C. M., Whetton, A. D., Wong, S., Martin, B. R., and Houselay, M. D., 1985, *Biochem. J.* **228:**593–603.

Higashijima, T., Ferguson, K. M., Smigel, M. D., and Gilman, A. G., 1987a, *J. Biol. Chem.* **262:**757–761.

Higashijima, T., Ferguson, K. M., Sternweis, P. C., Ross, E. M., Smigel, M. D., and Gilman, A. G., 1987b, *J. Biol. Chem.* **262:**752–756.

Hildebrandt, J. D., Codina, J., Rosenthal, W., Birnbaumer, L., Neer, E. J., Yamazaki, A., and Bitensky, M. W., 1985, *J. Biol. Chem.* **260:**14867–14872.

Hiratskua, T., and Uchida, K., 1980, *J. Biol. Chem.* **255:**1437–1448.

Ho, Y.-K., and Fung, B. K.-K., 1984, *J. Biol. Chem.* **259:**6694–6699.

Jelsema, C. L., and Axelrod, J., 1987, *Proc. Natl. Acad. Sci. USA* **84:**3623–3627.

Johnson, R. M., Connelly, P. A., Sisk, R. B., Pobiner, B. F., Hewlett, E. L., and Garrison, J. C., 1986, *Proc. Natl. Acad. Sci. USA* **83:**2032–2036.

Kobilka, B. K., Matsui, H., Kobilka, T. S., Yang-Feng, T. L., Francke, U., Caron, M. G., Lefkowitz, R. J., and Regan, J. W., 1987, *Science* **238:**650–656.

Kohnken, R. E., Eadie, D. M., Revzin, A., and McConnell, D. G., 1981, *J. Biol. Chem.* **256:**12502–12509.

Koland, J. G., and Cerione, R. A., 1988, *J. Biol. Chem.* **263:**2230–2237.

Kubo, T., Fukuda, K., Mikami, A., Maeda, A., Takahashi, H., Mishina, M., Haga, T., Haga, K., Ichiyama, A., Kangawa, K., Kojima, M., Matsuo, H., Hirose, T., and Numa, S., 1986a, *Nature* **321:**75–79.

Kubo, T., Maeda, A., Sugimoto, K., Akiba, I., Mikami, A., Takahashi, H., Haga, T., Haga, K., Ichiyama, A., Kangawa, K., Matsuo, H., Hirose, T., and Numa, S., 1986b, *FEBS Lett.* **209:**367–372.

Laws, W. R., and Shore, J. D., 1978, *J. Biol. Chem.* **253:**8593–8597.

Lefkowitz, R. J., and Caron, M. G., 1988, *J. Biol. Chem.* **263:**4993–4996.

Logothetis, D. E., Kurachi, Y., Galper, J., Neer, E. J., and Clapham, D. E., 1987, *Nature* **325**:321–326.

Luttrell, L. M., Hewlett, E. L., Romero, G., and Rogol, A. D., 1988, *J. Biol. Chem.* **263**:6134–6141.

May, D. C., Ross, E. M., Gilman, A. G., and Smigel, M. D., 1984, *J. Biol. Chem.* **260**:15829–15833.

Neer, E. J., and Clapham, D. E., 1988, *Nature* **333**:129–134.

Petruzelli, L., Herrera, R., and Rosen, O. M., 1984, *Proc. Natl. Acad. Sci. USA* **81**:3327–3331.

Pfeuffer, T., Gaugler, B., and Metzger, H., 1983, *FEBS Lett.* **164**:154–160.

Phillips, W. J., and Cerione, R. A., 1988, *J. Biol. Chem.* **263**:15498–15505.

Samana, J.-P., Zeppezauer, E., Biellmann, J.-F., and Branden, C.-I., 1977, *Eur. J. Biochem.* **81**:403–409.

Schechter, Y., Hernaez, L., Schlessinger, J., and Cuatrecasas, P., 1979, *Nature* **278**:835–838.

Schlessinger, J., Schechter, Y., Willingham, M. C., and Pastan, I., 1978, *Proc. Natl. Acad. Sci. USA* **75**:2659–2663.

Schreiber, A. B., Lax, I., Yarden, Y., Eshhar, Z., and Schlessinger, J., 1981, *Proc. Natl. Acad. Sci. USA* **78**:7535–7539.

Schreiber, A. B., Libermann, T. A., Lax, I., Yarden, Y., and Schlessinger, J., 1983, *J. Biol. Chem.* **258**:846–853.

Shia, M. A., Rubin, J. B., and Pilch, P. F., 1983, *J. Biol. Chem.* **258**:14450–14455.

Smigel, M. D., 1986, *J. Biol. Chem.* **261**:1976–1982.

Smith, C. D., Cox, C. C., and Snyderman, R., 1986, *Science* **232**:97–100.

Sternweis, P. C., and Gilman, A. G., 1982, *Proc. Natl. Acad. Sci. USA* **79**:4888–4891.

Stryer, L., 1978, *Annu. Rev. Biochem.* **47**:819–846.

Stryer, L., Hurley, J. B., and Fung, B. K.-K., 1981, *Curr. Top. Membr. Transp.* **15**:93–108.

Timasheff, S. N., 1970, in *The Enzymes,* Vol. 2 (P. Boyer, ed.), Academic Press, New York, pp. 418–430.

Ullrich, A., Coussens, L., Hayflick, J. S., Dull, T. J., Gray, A., Tam, A. W., Lee, J., Yarden, Y., Libermann, T. A., Schlessinger, J., Downward, J., Mayes, E. L. V., Whittle, N., Watefield, M. D., and Seeburg, P. H., 1984, *Nature* **309**:418–425.

Ullrich, A., Bell, J. R., Chen, E. Y., Herrara, R., Petruzelli, L. M., Dull, T. J., Gray, A., Coussens, L., Liao, Y.-C., Tsubokawa, M., Mason, A., Seeburg, P. H., Grunfeld, C., Rosen, O. M., and Ramanchandran, J., 1985, *Nature* **313**:756–761.

Wallace, M. A., and Fain, J. N., 1985, *J. Biol. Chem.* **260**:9527–9530.

Yarden, Y., and Schlessinger, J., 1987a, *Biochemistry* **26**:1434–1442.

Yarden, Y., and Schlessinger, J., 1987b, *Biochemistry* **26**:1443–1481.

Yatani, A., Codina, J., Imoto, Y., Reeves, J. P., Birnbaumer, L., and Brown, A. M., 1987, *Science* **238**:1288–1291.

Zidovetzki, R., Yarden, Y., Schlessinger, J., and Jovin, T. M., 1981, *Proc. Natl. Acad. Sci. USA* **78**:6981–6985.

Chapter 6

Analysis of Ligand Binding and Cross-Linking of Receptors in Solution and on Cell Surfaces
Immunoglobulin E as a Model Receptor

Jon Erickson, Richard Posner, Byron Goldstein, David Holowka, and Barbara Baird

1. INTRODUCTION

Fluorescence measurements are among the most powerful methods for investigating structure and structural changes on cell surfaces. The basic strength of this method is that nanomolar concentrations of fluorophores can be detected in the presence of a high level of background noise provided by the cells. An ideal fluorescent probe absorbs and emits at wavelengths not in common with the cellular components, has a high quantum yield, and can be placed specifically into a location such that it is sensitive to the structural aspect of interest. Recently there has been a great expansion in the commercial availability of fluorescent probes with a broad range of fluorescent properties and reactive groups for conjugation (e.g., Haugland, 1989). The requirement for specific placement on a macromolecular/cellular complex

JON ERICKSON, RICHARD POSNER, DAVID HOLOWKA, and BARBARA BAIRD • Department of Chemistry, Cornell University, Ithaca, New York 14853.   BYRON GOLDSTEIN • Theoretical Division, Los Alamos National Laboratory, Los Alamos, New Mexico 87545.   *Present address for J.E.*: Pierre A. Fish Laboratory, Department of Pharmacology, Cornell University, Ithaca, New York 14853.

remains the most challenging experimentally. In this regard, specific ligands and specific monoclonal antibodies that can be fluorescently modified are valuable reagents. In this chapter we describe our use of quantitative fluorescence measurements to investigate the binding properties of cell surface receptors.

A process of fundamental importance in cell biology is the binding of specific ligands to cell surface receptors that leads to signal transduction across the plasma membrane and causes a cellular response. In some cases, ligand binding to a single receptor is not sufficient, and cross-linking by ligand of two or more receptors is required to stimulate the response. This phenomenon is well illustrated in immunological systems with cell surface-associated immunoglobulin and antigen ligands. For example, cross-linking of surface immunoglobulin on B lymphocytes by certain antigens is a primary signal for proliferation and differentiation into antibody-secreting cells. This chapter focuses on another example: antigen-mediated cross-linking of immunoglobulin E (IgE)–receptor complexes on mast cells and basophils which leads to degranulation in an organism's allergic response (Ishizaka *et al.*, 1981). The critical molecular features of the cross-linking events that lead to a biological response in these cellular systems are not well understood. In order to address this problem systematically we have developed a model experimental system based on ligand binding to IgE in solution and to IgE–receptor complexes on the cell surface.

Physiologically, the allergic degranulation response mediated by the IgE antibody is initiated by bivalent or multivalent ligands (antigens) that can cluster IgE–receptor complexes that ordinarily are randomly dispersed on the cell surface. IgE is a soluble molecule, and it binds within its Fc segment to specific cell surface receptors (designated $Fc_{\epsilon}RI$) with high affinity ($K_a \geq 10^{10}$ M^{-1}; Metzger *et al.*, 1986). Each receptor binds one IgE antibody. Each IgE has two Fab segments that project up from the cell surface when IgE is bound to its receptor (Baird and Holowka, 1988), and each of these Fab segments possesses a binding site that recognizes the molecular determinants (haptens) on the antigen molecules. *In vivo,* mast cells have on their surface $\sim 10^5$ IgE receptors that bind serum IgE. The IgE is produced by lymphocytes in response to a variety of foreign antigens and provides each cell with a broad spectrum of affinities and recognition capabilities. Cross-linking <10% of the surface IgE receptors results in a full exocytotic response (Fewtrell, 1985).

Rat basophilic leukemia (RBL) cells are a tumor cell line derived from rat mucosal mast cells that have $\sim 3 \times 10^5$ high-affinity receptors for IgE per cell and that undergo IgE receptor-mediated degranulation (Barsumian *et al.*, 1981). These cells are particularly attractive to the experimentalist because they can be grown in tissue culture in the absence of IgE, and the receptors can be occupied to different extents with a variety of monoclonal IgE antibodies selected for specificity and affinity for a given ligand. In this manner the specificity and surface density of IgE receptors can be controlled. The ligands also can be selected or synthesized according to desired valency and structural properties. As illustrated in this chapter, this feature of singular molecular specificity in such an experimental system permits a detailed examination of the binding and cross-linking of antibodies with simple ligands. An important feature developed for this system is the use of fluorescently modified IgE which allows direct monitoring of ligand binding.

2. THE MODEL SYSTEM

A schematic representation of the relevant features of the experimental system is shown in Fig. 1. As depicted, the monoclonal IgE employed in these studies recognizes the 2,4-dinitrophenyl (DNP) hapten. This antibody has been chemically modified with fluorescein isothiocyanate (FITC). The DNP haptenic group has been conjugated to small peptides or single amino acids to provide a number of structurally well-defined ligands (Kane *et al.*, 1986). In this chapter, we will focus on the equilibrium and kinetic properties of three of these: the monovalent ligands, ϵ-DNP-L-lysine (DNP-lys) and DNP-aminocaproyl-L-tyrosine (DCT), and the bivalent ligand, N,N'-$(DCT)_2$-L-cystine [$(DCT)_2$-cys]. The chemical formulas of these ligands are provided in Fig. 2.

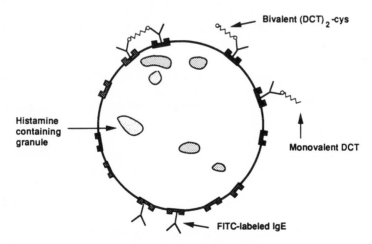

FIGURE 1. Schematic representation of an RBL cell and the interaction between receptor-bound FITC-anti-DNP-IgE and either monovalent DCT or bivalent $(DCT)_2$-cys.

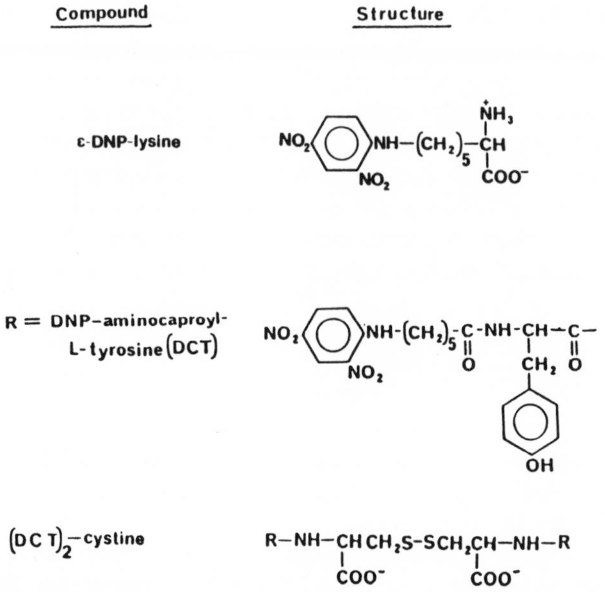

Compound	Structure
ε-DNP-lysine	
R = DNP-aminocaproyl- L-tyrosine (DCT)	
(DCT)$_2$–cystine	R–NH–CH CH$_2$S–SCH$_2$CH–NH–R

FIGURE 2. Chemical formulas of the DNP ligands used in these studies.

The basis of the binding assay used in these studies is the quenching of fluorescein fluorescence that accompanies the binding of DNP groups to the anti-DNP combining sites of FITC-IgE (Erickson *et al.*, 1986). A major advantage of using fluorescence quenching as a measure of antibody site occupation is the ability to monitor in real time the binding and cross-linking of ligand–IgE mixtures when the antibody is in solution or, more importantly, bound to its cell surface receptor. The fluorescein fluorescence excitation and emission maxima are ~490 and ~520 nm, respectively, which are longer wavelengths than those corresponding to most endogenous fluorescence. Hence, background signals are minimal. The sensitivity of the fluorescence method is such that the association of nanomolar concentrations of ligand and IgE can be easily detected. This aspect of the model system allows the study of the dynamics of ligand–antibody association rates even for the case where the bimolecular rate constant approaches a diffusion-limited value. This provides a method for studying these dynamics in solution and with antibody bound to living cells without the use of fast reaction techniques.

Incubation of the RBL cells with the DNP-specific IgE renders the cells responsive to multivalent DNP-containing ligands that cross-link two or more IgE–receptor complexes together. The degranulation response can be conveniently measured by allowing the cells to take up [^{3}H]serotonin which is then sequestered within secretory granules. The percentage of total cellular [^{3}H]serotonin released into the surrounding medium following the addition of a particular antigen serves as a measure of the efficacy of a particular stimulus. The physical factors that control the cross-linking events and thereby determine to some degree this efficacy probably arise from several sources including the structures and flexibilities of the antigen and receptor-bound IgE and the density and mobility of the receptors in the plasma membrane.

3. FLUORESCENCE BINDING METHOD

This method has been described in detail previously (Erickson et al., 1986). For our studies, monoclonal anti-DNP IgE (H1 26.82; Liu et al., 1980) is reacted with FITC, and we observe quenching of the fluorescein fluorescence that accompanies the binding of DNP to this modified IgE. The quenching is probably due to resonance energy transfer, and the maximal quenching observed (20%–30%, depending on the ligand) does not depend strongly on the stoichiometry of the FITC modification (4 to 12 fluoresceins per IgE antibody). Fluorescence titrations are carried out with a steady-state spectrofluorometer operated in ratio mode. Typically, 2 ml of a solution containing FITC-IgE or RBL cells saturated with FITC-IgE is stirred continuously in a thermostatically controlled cuvette holder. To these samples, solutions containing DNP ligands are added in microliter amounts with microcapillary tubes. The fluorescence signal can be averaged for a time sufficient to verify that a constant signal is obtained in the case of equilibrium experiments, or that acquisition periods are sufficiently long to give a desired signal-to-noise ratio in the case of kinetic measurements.

The FITC-IgE is also trace-labeled with ^{125}I for purposes of monitoring the total concentration of antibody binding sites (i.e., total number of Fab sites; X_T) in any experimental sample. For these determinations the specific activity of ^{125}I (cpm per mole sites) is obtained from a standardization titration with DNP-lys of that particular preparation of [^{125}I]-FITC-IgE (Erickson et al., 1986). In this standardization procedure, a high concentration of antibody combining sites ($[X]_T > 100$ nM) is titrated with DNP-lys. The initial part of the curve, corresponding to stoichiometric binding, is linear, and this is extrapolated to intersect the asymptote to the curve at a large excess of DNP where several additions of ligand indicate that no further quenching of the fluorescence signal is occurring. The binding site concentration is then the concentration of DNP-lys at which the two extrapolated lines intersect, and this value is used to calculate a specific activity for the $[^{125}I]$-FITC-IgE in terms of cpm per mole of binding sites.

4. DATA ANALYSIS

We present general derivations for the binding of bivalent ligands, and show how these can be simplified for monovalent ligand binding and other special cases. A minimal model for treating the equilibrium and kinetic data, shown in Fig. 3, considers the binding and cross-linking reactions between bivalent ligand and IgE sites in terms of the IgE antibody's two equivalent Fab binding sites (Erickson *et al.*, 1986). The concentration variables shown in Fig. 3 are defined as follows: [X], the concentration of free Fab sites; [Y], the concentration of Fab–$(DNP)_2$ complexes; [Z], the concentration of Fab–$(DNP)_2$–Fab complexes; and [C], the concentration of free bivalent ligand. Because the fractional quenching of FITC-IgE fluorescence is linearly related to the degree of Fab occupancy at all points in a titration curve (Erickson *et al.*, 1986), the level of fluorescence gives directly the proportions of free $(=[X])$ and bound $(=[Y] + 2\ [Z])$ Fab sites. The model reduces the number of bound states to two (Y and Z) by imposing the following simplifying assumptions: (1) The ligand cannot bridge two Fabs on the same IgE (i.e., no intramolecular cyclization). (2) Linear chains of IgE formed when ligands bind bivalently do not form rings (i.e., no intermolecular cyclization). (3) For the forward cross-linking step, the reactivity of the free ligand end in complex Y, or a free Fab end are unaffected by the size of the chain to which

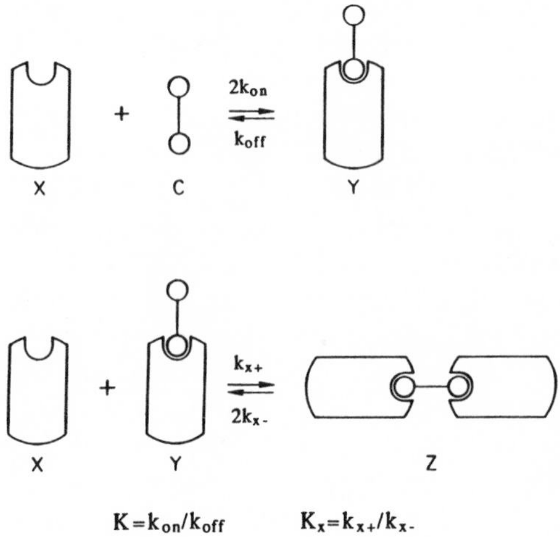

FIGURE 3. Reaction scheme defining binding and cross-linking reactions that can occur between bivalent ligand and Fab combining sites on IgE. Adapted from Erickson *et al.* (1986).

they may be attached. (4) Similarly for the complex Z, the intrinsic dissociation rate of one end of a ligand is independent of the size of the chain in which it is incorporated. Statements 3 and 4 constitute the "equivalent site hypothesis" (Flory, 1953).

The reactions shown in Fig. 3 define two equilibrium constants and therefore four rate constants: $K = k_{on}/k_{off}$ is the binding constant for the association of a free Fab binding site with a hapten site on a free bivalent ligand, and $K_x = k_{x+}/k_{x-}$ is the cross-linking constant for the interaction of a free Fab binding site with the remaining free hapten binding site on a bivalent ligand–Fab complex. Equations are derived in the following sections in order to express the experimentally observed fraction of bound Fab sites as a function of the independent variables: total concentration of DNP ($[D]_T$) in the case of the equilibrium titrations, or elapsed time in the case of the kinetic experiments.

4.1. Equilibrium Binding of Monovalent and Bivalent Ligands

Application of the minimal model (Fig. 3) to experimental data in order to obtain estimates for K and K_x requires an expression for the observed value of q, the fraction of Fab binding sites bound to DNP, as a function of known quantities: $[D]_T$, the total concentration of DNP groups, and $[X]_T$, the total concentration of Fab binding sites. At equilibrium (Dembo and Goldstein, 1978a),

$$Y = 2K[C][X] \tag{1}$$

$$Z = K_x K[C][X]^2 \tag{2}$$

In addition, conservation of Fab fragments and ligand requires that

$$[X]_T = [X] + 2K[C][X] + 2KK_x[C][X]^2 \tag{3}$$

$$[C]_T = [C] + 2K[C][X] + KK_x[C][X]^2 \tag{4}$$

$$[X] = [X]_T(1 - q) \tag{5}$$

where $[C]_T = [D]_T/2$ is the total bivalent ligand concentration. Since

$$q = \frac{\text{Fab sites bound}}{\text{total Fab sites}} = \frac{[Y] + 2[Z]}{[X] + [Y] + 2[Z]} \tag{6}$$

q can also be written in terms of $[D]_T$ and $[X]$ by substitution of Eqs. (1) and (2) into Eq. (6) and then using Eq. (4) to eliminate $[C]$,

$$q = \frac{K[D]_T + KK_x[D]_T[X]}{1 + 2K[X] + KK_x[X]^2 + K[D]_T + KK_x[D]_T[X]} \tag{7}$$

q can be expressed in terms of $[X]_T$ rather than $[X]$ by substitution of Eq. (5), then Eq. (7) becomes cubic in q and may be solved numerically in order to provide best values of the equilibrium parameters K and K_x. The use of nondimensional variables allows binding experiments at different Fab concentration to be fit simultaneously (Erickson *et al.*, 1986).

Equation (7) can be rewritten for the case of monovalent DNP ligand binding to IgE combining sites by setting $K_x = K$. This substitution into Eq. (7) together with Eq. (5) yields

$$q = 1/2\{([D]_T/[X]_T + 1/K[X]_T + 1)$$
$$- (([D]_T/[X]_T + 1/K[X]_T + 1)^2 - 4[D]_T/[X]_T)^{1/2}\} \quad (8)$$

4.2. Kinetics of Ligand Binding and Cross-Linking

4.2.1. Monovalent Ligand Binding Kinetics

Data analysis is carried out assuming that the reaction between monovalent ligand and surface receptor can be described by a general one-step bimolecular reaction scheme that is consistent with the general scheme of Fig. 3 (Erickson *et al.*, 1987). Consider the reaction

$$L + X \underset{k_{off}}{\overset{k_{on}}{\rightleftharpoons}} L\text{-}X$$

where L and X represent the free monovalent DNP ligand and unbound Fab sites, respectively, and $[L\text{-}X]$ is the concentration of bound complex with all quantities written as moles per liter. The experimentally observable quantity, the decay of fluorescence that accompanies the binding of ligand and therefore the disappearance of unoccupied Fab, may be written as

$$-d[X]/dt = k_{on}[L][X] - k_{off}[L\text{-}X] \quad (9)$$

The conservation expressions for total DNP and Fab are

$$[L]_T = [L] + [L\text{-}X] \quad (10)$$

$$[X]_T = [X] + [L\text{-}X] \quad (11)$$

Substitution of Eqs. (10) and (11) into Eq. (9) leads to

$$-d[X]/\{(-k_{off}[X]_T) + (k_{off} + k_{on}([L]_T - [X]_T))[X] + k_{on}[X]^2\} = dt \quad (12)$$

where the variables have been separated. If we assume k_{on} and k_{off} are constant during the course of the binding, integration from $[X] = [X]_T$ at $t = 0$ to $[X] = [X(t)]$ at time t followed by rearrangement yields

$$[X(t)]/[X]_T = \{b(\alpha e^{t\beta} - 1) + \beta(\alpha e^{t\beta} + 1)\}/\{2c(1 - \alpha e^{t\beta})[X]_T\}$$

$$(13)$$

where

$a = k_{off}[X]_T$
$b = k_{on}([X]_T - [L]_T) - k_{off}$
$c = -k_{on}$
$\alpha = \{2c[X]_T + b - \beta\}/\{2c[X]_T + b + \beta\}$
$\beta = (b^2 - 4ac)^{1/2}$

Normalized data sets can be fit to Eq. (13) to extract k_{on}. Knowledge of $[X]_T$ from the specific activity of the IgE and $[L]_T$ from accurate measurement of DNP in the ligand stock solutions also allows these parameters to be entered and held fixed for the computer fitting. Finally, since the association constant K for DCT and cell-bound $[^{125}I]$-FITC-IgE is known, k_{on}/K_a can be substituted for k_{off} in Eq. (13) reducing the number of freely varying parameters to one.

4.2.2. Bivalent Ligand Binding Kinetics

In general, the binding of bivalent ligands to bivalent receptors leads to the formation of ligand–receptor chains and rings. Thus, at any time there can be many different size chains and rings present at different concentrations. The distribution of such aggregates will continue to change until the equilibrium distribution is reached. One possible way to describe the kinetics of such binding is to write a differential equation for each different size chain and ring that can possibly form. However, Perelson and DeLisi (1980) showed that under the assumptions outlined at the beginning of Section 4, an equivalent description can be obtained by writing differential equations to describe the possible states of the ligand. There are only three of these, C, Y, and Z, corresponding to ligands with both ends free, one end free and one end bound, or both ends bound, respectively. Since the total amount of ligand is conserved, only two differential equations are required, because if two ligand states are known the third can be calculated from the conservation law. Perelson and DeLisi (1980) chose to write differential equations to describe the evolution in terms of the two bound states of the ligand. In our notation these equations become

$$-d[Y]/dt = -2k_{on}[X][C] + k_{off}[Y] + k_{x+}[X][Y] - 2k_{x-}[Z] \quad (14a)$$

$$-d[Z]/dt = -k_{x+}[X][Y] + 2k_{x-}[Z] \quad (14b)$$

In addition, we have the two conservation laws

$$[X]_T = [X] + [Y] + 2[Z] \qquad (14c)$$

$$[C]_T = [C] + [Y] + [Z] \qquad (14d)$$

Equations (14a) and (14b) can be used to eliminate $[X]$ and $[C]$ and obtain two coupled nonlinear differential equations for $[Y]$ and $[Z]$. These equations cannot be solved in closed form. In general, these equations must be solved numerically, although for certain parameter values approximate analytic solutions can be obtained.

4.2.3. Dissociation Kinetics

If ligand is initially in equilibrium with IgE and then experimental conditions are fixed such that the forward reactions in Fig. 3 are prevented from occurring (for example by blocking all free Fab sites), the model reduces to one describing the sequential release of $(DCT)_2$-cys from solution or cell surface-bound IgE. The dissociation steps are: (1) the breakup of a cross-link, when one Fab–DNP interaction in a complex Z is broken (described by k_{x-}) and (2) dissociation of the resulting monovalently bound ligand into bulk solution (described by k_{off}). Differential equations for the decay of the bound ligand states follow from Eqs. (14a) and (14b) when we set $[X] = 0$:

$$-d[Z]/dt = 2k_{x-}[Z] \qquad (15)$$

$$-d[Y]/dt = k_{off}[Y] - 2k_{x-}[Z] \qquad (16)$$

Integration of Eq. (15) gives directly

$$[Z(t)] = [Z(0)]\exp(-2k_{x-}t) \qquad (17)$$

where $[Z(0)] = [Z(t)]$ at $t = 0$. Substitution into Eq. (17) followed by corresponding integration yields

$$[Y(t)] = \exp(-k_{off}t)\{(2k_{x-}[Z(0)]\exp((k_{off} - 2k_{x-})t) \\ - 2k_{x-}[Z(0)])/(k_{off} - 2k_{x-}) + [Y(0)]\} \qquad (18)$$

Equations (15) and (16) can be combined to give an analytical expression for occupied Fab as a function of time since $[Fab(t)]_{bound} = [Y(t)] + 2[Z(t)]$. Rearrangement leads to

$$[Fab(t)]_{bound} = \exp(-k_{off}t)\{[Y(0)] - ((2k_{x-}[Z(0)])/(k_{off} - 2k_{x-}))\} \\ + \exp(-2k_{x-}t)\{2[Z(0)] + (2k_{x-}[Z(0)])/(k_{off} - 2k_{x-})\} \qquad (19)$$

Equations (19) defines precisely the two-exponential decay of the bound Fab states expected from two-step unidirectional dissociation in the scheme in

Fig. 3. Although we have assumed that rings cannot form, it can be shown that Eq. (19) is valid even if rings are present, if the rate constant for the breaking of cross-links, k_{x-}, is the same for cross-links in chains and rings.

Note that when $k_{x-} = k_{off}$, the entire course of dissociation will be well described by a single exponential

$$[Fab(t)]_{bound} = \exp(-k_{off}t)\{[Y(0)] + 2[Z(0)]\} \tag{20}$$

Similarly, if only monovalent or monovalently bound ligand is considered (i.e., $[Z(0)] = 0$), Eq. (20) further reduces to

$$[Fab(t)]_{bound} = \exp(-k_{off}t)[Y(0)] \tag{21}$$

Previously determined physical constants for this system assist in the data analysis by providing initial estimates of parameter values as well as cross-checks for the results provided by the computer fitting of the dissociation data.

5. EXPERIMENTAL RESULTS

5.1. Tests of the Binding Method

These experiments illustrate the validity of the fluorescein fluorescence quenching method by providing values for an affinity constant, K, that are comparable to those obtained for the same ligand receptor pair using previously established methods. Figures 4a and b compare the binding of DNP-lys to monoclonal anti-DNP IgE (H1 26.82; Liu et al., 1980) in solution as measured by two different fluorescence quenching methods. In Fig. 4a binding is assessed by monitoring the quenching of endogenous IgE tryptophan fluorescence as a function of DNP-lys titrated into a solution of unmodified anti-DNP IgE (Eisen and McGuigan, 1968). Since the fractional quenching of tryptophan fluorescence in the monoclonal antibody should be the same as the fraction of combining sites occupied by the monovalent ligand, this latter quantity can be obtained directly from the raw data and plotted as a function of total DNP titrated into the solution. Nonlinear least-squares fitting of the data according to Eq. (8) as shown in Fig. 4a yields $K = 1.2 \times 10^8$ M^{-1}, and this agrees very well with the value previously determined for this monoclonal antibody in solution with an equilibrium dialysis method ($K = 1.4 \times 10^8$ M^{-1}; Liu et al., 1980).

Shown in Fig. 4b are data from a representative experiment in which fluorescein quenching of FITC-IgE was monitored as a function of DNP-lys concentration. These data are plotted as the fraction of the combining sites occupied in an analogous fashion to the data in Fig. 4a, and Eq. (8) fits the data with $K = 1.9 \times 10^8$ M^{-1} (Table I). Titration with DNP-lys of Fab

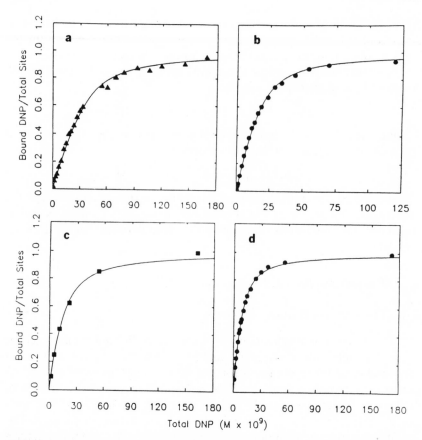

FIGURE 4. Equilibrium binding of DNP-lys to IgE in solution (a, 17 nM IgE and b, 8.5 nM IgE) or on cells (c, 5.5 nM IgE and d, 3.6 nM IgE) as measured by quenching of tryptophan fluorescence (a), quenching of fluorescein fluorescence (b and d), or association of [^{3}H]-DNP-lys (c). The curves are fit with Eq. (8) and the following values of K: (a) 1.2×10^8 M^{-1}; (b) 1.9×10^8 M^{-1}; (c) 1.2×10^8 M^{-1}; (d) 2.4×10^8 M^{-1}. Adapted from Erickson *et al.* (1986).

fragments prepared from FITC-anti-DNP IgE by proteolytic digestion (Erickson *et al.*, 1986) yields a curve similar to that shown in Fig. 4b, and the data are well fitted by Eq. (8) and $K = 1.7 \times 10^8$ M^{-1} (Table I).

Binding of DNP-lys also can be assessed when the FITC-IgE is bound to its high-affinity receptor on the surface of RBL cells. In this situation, the tryptophan quenching method is not practical due to the presence of other cellular proteins and, correspondingly, an overwhelming level of background tryptophan fluorescence. Therefore, comparison is made to the binding of [^{3}H]-DNP-lys to unmodified anti-DNP IgE under similar conditions.

TABLE I. Equilibrium Constants for DNP Ligands Binding to Anti-DNP IgE or Its Fab Fragment in Solution or to Anti-DNP IgE Bound to Receptors on RBL Cells[a]

Anti-DNP	Ligand	$K \, (M^{-1} \times 10^{-9})$	$K_x \, (M^{-1} \times 10^{-7})$
Fab (solution)	DNP-lys	0.17	—
IgE (solution)	DNP-lys	0.19	—
IgE (cell)	DNP-lys	0.24	—
Fab (solution)	DCT	2.1	—
Fab (solution)	$(DCT)_2$-cys	2.2	3.0
IgE (solution)	DCT	1.8	—
IgE (solution)	$(DCT)_2$-cys	1.6	7.8
IgE (solution)	$(DCT)_2$-cys (red)[b]	1.7	—
IgE (cell)	DCT	1.5	—

[a] From Erikson *et al.* (1986).
[b] $(DCT)_2$-cys reduced to monovalent DCT-cysteine in the presence of 1 mM dithiothreitol.

Figure 4c shows data from a representative experiment in which RBL cells were saturated with anti-DNP IgE, and binding of [^{3}H]-DNP-lys was assessed. The data are well fitted by Eq. (8) to yield $K = 1.2 \times 10^8 \, M^{-1}$. The number of molecules of [^{3}H]-DNP-lys bound at saturation in Fig. 4c was determined to be 3.2×10^5/cell, consistent with the general observation of 1.5–3.0×10^5 molecules of [^{125}I]-IgE binding per RBL cell at saturation.

Figure 4d shows the results from a representative experiment in which DNP-lys binding to FITC-IgE on the cell surface was assessed by monitoring the quenching of FITC. A good fit to these data is obtained with Eq. (8) and $K = 2.4 \times 10^8 \, M^{-1}$ (Table I). [^{125}I]-FITC-IgE bound per cell ranged from 1 to 3×10^5 in this and similar experiments, and no correlation between the value of K and the number of IgE molecules bound per cell was observed, as expected. The conclusion drawn from all of the results is that the fluorescein quenching method provides an accurate value for K for both cell-bound and solution anti-DNP IgE.

5.2. Equilibrium Binding Experiments

These experiments, which have been described in detail previously (Erickson *et al.*, 1986), examine monovalent (DCT) and bivalent [$(DCT)_2$-cys] ligands (Fig. 2) binding to monovalent (Fab fragment) and bivalent (IgE, expressed in terms of Fab sites) receptors. Table I presents the intrinsic and cross-linking equilibrium constants (K and K_x, respectively; see Fig. 3) obtained from several experiments carried out over a range of concentrations of Fab sites for IgE in solution and confined to the cell surface.

5.2.1. Binding to Receptors in Solution

For DCT binding to Fab fragments of anti-DNP IgE in solution the data are well fitted by Eq. (8) and yield an intrinsic affinity constant about an order of magnitude greater than that for DNP-lys (Table I). This difference suggests that the tyrosine residue contained in DCT contributes to the binding interaction with this monoclonal anti-DNP IgE. The binding of the bivalent ligand (DCT)$_2$-cys to anti-DNP Fab cannot be fit with a single intrinsic affinity constant but is well fitted by Eq. (7) which includes a second affinity constant, K_x, that describes the cross-linking reaction. The similarity in the values of K obtained for DCT and (DCT)$_2$-cys suggests that the DCT moiety retains the same intrinsic affinity after incorporation into (DCT)$_2$-cys. The 100-fold lower value of K_x for (DCT)$_2$-cys probably reflects steric restrictions on the unbound end of the ligand due to the close proximity of the much larger antibody structure, which results in a reduced DNP/Fab affinity. A simple model based on flexible haptens and spherical Fab has been developed (Dembo and Goldstein, 1978a) that predicts the dependence of the ratio K_x/K on three structural parameters: radius of the Fab, antibody combining site depth, and effective length of the bivalent ligand. Application of this theory and choice of appropriate values for the other parameters indicates that the monoclonal antibody H1 26.82 has a relatively deep binding site of about 20 Å which is consistent with its observed high affinity for the monovalent ligand DCT (cf. Carson and Metzger, 1974). The symmetrical nature of (DCT)$_2$-cys can be investigated further since the molecule has a disulfide bond that can be subjected to reducing reagents to yield the monovalent analogue, DCT-cysteine (see Fig. 2). Consistent with the results cited above, the binding data for IgE and (DCT)$_2$-cys after reduction are well fitted by a single equilibrium constant having a similar value to that obtained for intact (DCT)$_2$-cys and for DCT (Table I).

The equilibrium binding results obtained with the monovalent anti-DNP Fab fragments can be compared with those for intact, bivalent anti-DNP IgE, where binding and cross-linking by bivalent ligand may form chains and cyclic structures. The data for DCT and (DCT)$_2$-cys are well fitted by Eq. (8) and (7), respectively, and representative data for (DCT)$_2$-cys are included in Fig. 5 (open triangles). As shown in Table I, values of K obtained for DCT and (DCT)$_2$-cys binding to IgE in solution are similar to those obtained for Fab fragments. The value of K_x obtained for (DCT)$_2$-cys binding to IgE is somewhat larger than that for binding to Fab fragments, and this may reflect additional ligand states than those depicted in Fig. 3 (e.g., ligands incorporated into cyclic complexes of chains). As discussed in Section 4, linear chains are included in the model if the "equivalent site" hypothesis is valid. However, to consider properly the possibility of cyclic structures re-

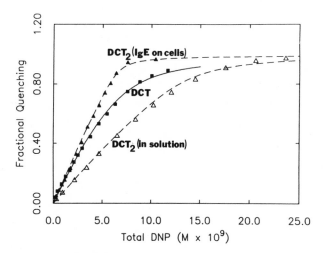

FIGURE 5. Comparison of (DCT)$_2$-cys binding to IgE in solution ($\triangle$; $K = 1.8 \times 10^9$ M^{-1}, K_x = 1.2×10^7 M^{-1}), (DCT)$_2$-cys binding to IgE on cells ($\blacktriangle$; $K = 2.1 \times 10^9$ M^{-1}, $K_x = 1.5 \times 10^{10}$ M^{-1}), and DCT binding to IgE on cells ($\blacksquare$; $K = 1.5 \times 10^9$ M^{-1}).

The total concentration of IgE combining sites in each experiment was 7 nM. The data are fit with Eq. (8) for DCT and with Eq. (7) for (DCT)$_2$-cys. As shown, (DCT)$_2$-cys binding is weaker than DCT for IgE in solution but stronger than DCT when the IgE is confined to a cell surface, because binding of the second end of the bivalent ligand is facilitated by high local concentration of IgE on the cells. As discussed in the text, equilibrium for bivalent ligand binding to IgE on cells is not well defined, since cellular processes may alter conditions.

quires incorporation of at least two additional affinity constants that correspond to intramolecular cross-linking and to cyclization of ligand–antibody chains (Dembo and Goldstein, 1978b). A possible explanation for the larger value of K_x obtained for IgE is that this value represents an average that includes some chain and cyclic structures. Others have included cyclization equilibrium constants in more detailed analyses with somewhat different ligands and antibodies (Schweitzer-Stenner *et al.*, 1987).

5.2.2. Binding to Receptors on the Cell Surface

As stated above, a major advantage of the fluorescein quenching method is that it allows measurement and analysis of the binding to and cross-linking of cell surface receptor-bound IgE by DNP ligands. As shown in Fig. 5, titration of RBL cells saturated with FITC-anti-DNP IgE with the monovalent ligand, DCT (squares), results in a simple binding curve that can be fit with Eq. (8). The average K for this case is similar to the average value obtained for DCT with anti-DNP IgE and Fab in solution (Table I).

Figure 5 also includes data from a titration with (DCT)$_2$-cys of cell-bound FITC-IgE (solid triangles). A comparison of the two curves indicates that a large fraction of the (DCT)$_2$-cys must be binding bivalently since the data points for this ligand fall on the straight line $q = [D]_T/[X]_T$ for most of the range where $[D]_T < [X]_T$. These data for the bivalent ligand and cell-bound IgE can be fit satisfactorily with Eq. (7) to yield a value for $K_x \sim$ 100-fold greater than the K_x obtained for the same ligand and IgE in solution (Fig. 5, compare solid and open triangles; Table I). This large increase in the apparent three-dimensional K_x reflects the confinement of the IgE molecules to the cell surface resulting in greatly increased local concentration and corresponding enhancement of the cross-linking step. Again, the nature of the cross-linked species and their size distribution are undetermined, and it is possible that both linear and cyclic oligomers of IgE contribute to the cross-linked population (Dembo and Goldstein, 1978b) and to the resulting average value of K_x obtained.

It should be noted that for reactions occurring on cell surfaces the intrinsic cross-linking constant is two-dimensional having units of cm^2, i.e., (surface concentration)$^{-1}$. The apparent three-dimensional cross-linking constant is related to the two-dimensional cross-linking constant as: $K_x^{2D} = K_x^{3D} A\rho/6.02 \times 10^{20}$, where A is the average surface area of the cell in cm^2, ρ is the solution concentration of cells in cells/ml, and K_x^{3D} is the apparent equilibrium cross-linking constant in M^{-1}. The surface area of an RBL cell is $\sim 5 \times 10^{-6}$ cm^2. In our experiment $\rho \sim 10^7$ cells/ml, and we found $K_x^{3D} = 1.5 \times 10^{10}$ M^{-1} so that $K_x^{2D} = 1.25 \times 10^{-9}$ cm^2. From K_x^{2D} we can determine the cell surface receptor density that is required to obtain significant cross-linking (Dembo *et al.*, 1979). The amount of cross-linking present at equilibrium increases as bivalent ligand concentration ([C]) increases, goes through a maximum at $[C] = 1/2K$, and then decreases to zero. For roughly half of the receptors to be involved in cross-links at the optimum ligand concentration ($=1/2K$) requires that $K_x^{2D}X_T \geq 1$, where X_T is the surface receptor density. For (DCT)$_2$-cys this means that $X_T \geq 8 \times 10^8$ receptors/cm^2 or 4×10^3 receptors/cell.

Additional uncertainty arises in considering ligand binding to and cross-linking of receptors on the surface of cells because the cells may begin to change in response to these binding events. If receptor properties that affect ligand binding are changing as the ligand binding proceeds, then the equilibrium state becomes undetermined even though the quantity of bound states appears to remain constant. In this regard it is known that binding of some bivalent and multivalent ligands to IgE–receptor complexes on RBL cells results in an immediate reduction in the free lateral diffusion of the receptors (Menon *et al.*, 1986b; Kane *et al.*, 1988) and, after several to many minutes, association of the receptors with the cytoskeletal matrix (Robertson *et al.,* 1986) and redistribution of the receptors into large clusters (patches) (Menon *et al.,* 1984). Endocytosis of the cross-linked receptors can also occur, although this phenomenon is detectable by fluorescence and can be prevented by reducing the temperature or by treating the cells with a combi- nation of azide and deoxyglucose (Menon *et al.,* 1986a). In the fluorescein quenching binding method with multivalent ligands and cell surface IgE we generally observe that the quenching reaches a steady level within several minutes (see Fig. 6), indicating that the net occupation of Fab binding sites is stable. However, redistribution of the cross-linked species probably contin- ues (see Section 5.3.2). Ignorance of the details of these events results in additional uncertainty of the meaning of the K_x derived from application of Eq. (7). The fact that cell systems are dynamic and responsive in nature underscores the importance of analyzing the *kinetics* of ligand binding.

5.3. Kinetic Binding Experiments

The sensitivity of the fluorescein quenching method is such that binding between nanomolar quantities of Fab and DNP can be accurately measured. Under these conditions of low concentrations, binding reactions between ligand and antibody proceed on a time scale of several seconds even though they are characterized by forward rate constants on the order of 10^7 M^{-1} sec^{-1} (Pecht and Lancet, 1977). This means that the binding can be followed without resorting to fast kinetic techniques that are difficult to apply to cellular systems.

5.3.1. Kinetics of Ligand Binding to Cell Surface Receptors

A typical trace illustrating the time course of the monovalent ligand DCT binding to cell surface IgE is shown in Fig. 6a. The data are well fitted by Eq. (13) which necessarily contains both the forward and reverse rate constants. The ability to make these kinetic measurements allows a test of the predicted functional dependence of the observed k_{on} on the surface den- sity of receptors (Berg and Purcell, 1977). This dependence may be written as

$$k_{on}(cm^3/cell \cdot sec) = 4\pi DaN\kappa_{on}/(4\pi Da + N\kappa_{on}) \qquad (22)$$

where k_{on} is the measured bimolecular forward rate constant for ligand bind- ing to cell surface receptors, and the ligand has a diffusion constant D in solution and binds to a cell that has radius a and possesses N receptors. κ_{on} is the intrinsic "reaction-limited" ligand–receptor forward rate constant. Equation (22) predicts that the rate constant for binding to the whole cell will be directly proportional to the number of receptors on the cell surface (N)

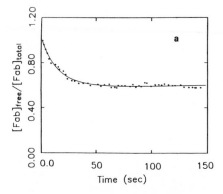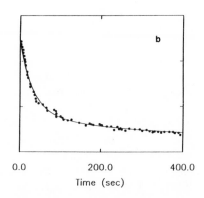

FIGURE 6. Kinetic curves for DCT (a) and $(DCT)_2$-cys (b) binding to IgE on cells. Data for DCT were obtained with a sample containing 2.3 nM IgE (2.4×10^6 cells/ml) and 4.2 nM DNP; the points are fit by Eq. (13) with $k_{on} = 2.6 \times 10^7$ M^{-1} sec^{-1} and $k_{off} = 1.9 \times 10^{-2}$ sec^{-1}. Data for $(DCT)_2$-cys were obtained with a sample containing 0.16 nM IgE (3.3×10^6 cells/ml) and 1.8 nM DNP; the points are fit by Eq. (14) with $k_{on} = 1.2 \times 10^7$ M^{-1} sec^{-1}, $k_{off} = 1.0 \times 10^{-2}$ sec^{-1}, $k_{x+} = 2.1 \times 10^7$ M^{-1} sec^{-1}, $k_{x-} = 4.0 \times 10^{-4}$ sec^{-1}. a and b from Erickson *et al.* (1987) and Erickson (1988), respectively.

when $N\kappa_{on} \ll 4\pi Da$. Conversely, when $N\kappa_{on} \gg 4\pi Da$, Eq. (22) predicts that a cell will approach a limit in its ligand adsorption efficiency, with a corresponding value for the observed rate constant of $\lim(k_{on}) = 4\pi Da$. The physical basis for the functional dependence of k_{on} on receptor density lies in the Brownian motion of the ligand at the cell surface. As a consequence of this motion, a single ligand encounters the cell surface many times before diffusing away and, in this way, increases its chances of successfully finding a receptor. For a receptor isolated from other receptors, the rate constant for ligand binding is $k_{on} = \kappa_{on}$. However, competition among the cell surface receptors as receptor density increases results in an observed decrease in the measured k_{on} per receptor. This predicted dependence on density of receptors has the important biological consequence that the cell captures a given ligand with nearly maximum efficiency when only a small percentage of its total surface area is occupied by receptors specific for that task (Berg and Purcell, 1977).

Figure 7 shows accumulated data from experiments performed under a variety of conditions of cell concentration, total IgE concentration ([Fab]$_{bulk}$), and IgE surface density (N). The observed k_{on} data are plotted as a function of N in order to illustrate the hyperbolic dependence predicted by Eq. (22). The best fit shown by the solid line was obtained by varying the composite parameter $4\pi Da$ and the intrinsic forward rate constant κ_{on}, and this fit yielded values of 4.96×10^{-8} cm^3/cell·sec and 1.86×10^{-13} cm^3/

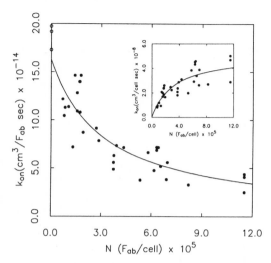

FIGURE 7. A plot of k_{on} (per receptor) versus number of receptors per cell (N) from several different experiments. The data are fit by Eq. (22) with $4\pi Da = 4.96 \times 10^{-8}$ cm^3 cell^{-1} sec^{-1} and κ_{on} = 18.6 $\times$ 10^{-14} cm^3 Fab^{-1} sec^{-1}. Inset: Same data plotted as k_{on}(per cell) = k_{on}(per receptor) $\times$ N versus N. The asymptotic value for this curve at high N is k_{on} = $4\pi Da$ = 4.96 $\times$ 10^{-8} cm^3 cell^{-1} sec^{-1}. From Erickson *et al.* (1987).

Fab · sec, respectively. The derived value for κ_{on} should be the same as the rate constant for DCT binding to isolated FITC-IgE in solution, and this is indeed observed experimentally (Fig. 7, $N = 0$). The plausible quantitative predictions of the model are further illustrated by choosing an approximate value for the DCT diffusion coefficient, $D = 10^{-5}$ cm^2/sec, and solving for a, the radius of the cell. This procedure yields an apparent cell radius of 4 μm which agrees with direct measurements of 4–5 μm (Isersky *et al.*, 1979).

The time course for binding of the bivalent ligand (DCT)$_2$-cys to FITC-IgE on cells is shown in Fig. 6b. The shape of the curve is different from that shown for monovalent DCT in Fig. 6a, reflecting the occurrence of cross-linking. The time needed for maximal binding of DNP as observed by fluorescence quenching is significantly longer for (DCT)$_2$-cys compared to DCT, and at least two phases are apparent. Analysis of (DCT)$_2$-cys binding to FITC-IgE on cells requires a minimum of four rate constants. More parameters may be involved since cross-linking of IgE receptors on the cell surface potentially results in additional complexity both in terms of the variety of cross-linked structures that may form and in terms of cellular responses that may alter the binding conditions. Figure 3 shows a minimal model which assumes that four rate constants adequately describe the distribution of bound and cross-linked states, and that the latter can be represented by a single type of cross-linked ligand species (Z). With this approximation the time course of quenching by (DCT)$_2$-cys is described by a system of two coupled differential equations [Eq. (14)]. We have used this equation to fit (DCT)$_2$-cys binding data with numerical methods, and good fits have yielded apparent values for k_{on}, k_{off}, k_{x+}, and k_{x-} that are consistent with the k_{on} and k_{off} obtained with DCT and with the K and K_x derived from equilibrium experiments (Table I) (Erickson, 1988). Although the data for (DCT)$_2$-cys binding can be adequately accounted for by Eq. (14) the apparent rate constants that are derived with this minimal model may reflect average values for a complex distribution of oligomeric receptor species.

5.3.2. Kinetics of Dissociation from Cell Surface Receptors

Kinetic analysis of unidirectional ligand dissociation requires consideration of only the reverse rate constants. Dissociation can be monitored experimentally if the measurements begin with some ligands bound and if the subsequent forward binding steps of ligand are prevented. The mathematical result for unidirectional dissociation within the minimal model is a simple sum of two exponentials describing the decay of the two bound states [Eq. (19)]. In our experiments, DNP ligand is first added to a sample containing FITC-IgE, and the fluorescence quenching that accompanies binding is monitored until it reaches a steady value. Under these conditions, net dissociation of DNP can be observed as an increase of fluorescence when the DNP-FITC-IgE binding steady state is perturbed. Net dissociation of bound ligand is initiated and rebinding inhibited by adding a high concentration of a competing reagent that blocks the Fab binding sites or that absorbs the released DNP ligand. We have used the following two methods to monitor the dissociation of (DCT)$_2$-cys from FITC-IgE: introduction into the solution of either an excess of unlabeled anti-DNP IgE or, alternatively, an excess of DNP-lys or DCT which quench the FITC-IgE fluorescence less than (DCT)$_2$-cys (20% for the monovalent ligands versus 30% for the bivalent ligand; Erickson *et al.*, 1986).

Figure 8 illustrates a typical trace from a cycle of fluorescence quenching caused by (DCT)$_2$-cys binding to cell surface FITC-IgE followed by fluorescence recovery that is initiated by the subsequent introduction of excess unlabeled anti-DNP IgE. The recovery curve obtained with monovalent DNP ligand as competitor is similar except that the final fluorescence level is about 67% lower due to quenching of the FITC-IgE by the competitor. With the minimal model the fluorescence recovery curve is fit with Eq. (19) or (21) to determine rate constants for bivalent or monovalent ligand dissociation, respectively. With this method, values must be determined for the fluorescence immediately after addition of the competitor and for the fluorescence after maximal recovery, and these may be treated as adjustable parameters in the data fitting routines (Goldstein *et al.*, 1989).

In dissociation experiments, use of unlabeled anti-DNP IgE as a competitor has the advantage that a greater fluorescence recovery accompanies dis-

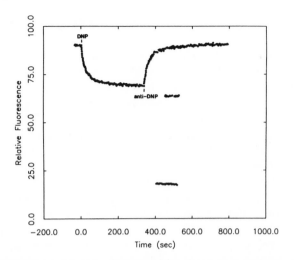

FIGURE 8. Fluorescence quenching of FITC-IgE bound to cells by $(DCT)_2$-cys (subsaturating amount) and subsequent recovery of fluorescence caused by the addition of excess unlabeled IgE to the suspension. Also shown are levels of fluorescence from an identical cell sample saturated with $(DCT)_2$-cys (maximal quenching) and a sample of cells with no bound FITC-IgE (background signal). From Erickson (1988).

sociation resulting in a better signal-to-noise ratio than that obtained with monovalent ligand as competitor. We have used unlabeled IgE as a competitor in order to monitor dissociation of monovalent ligand, DCT, from cell surface FITC-IgE (Goldstein *et al.,* 1989). This experimental situation could be exploited to examine the general problem of competition for monovalent ligand between receptors in solution and receptors on the cell surface. Such situations arise physiologically, for example, when B cells are stimulated by antigen binding to their surface immunoglobulin in the presence of secreted antibodies of the same specificity. As discussed in Section 5.3.1, dissociation of ligand from receptors on the cell surface should be slowed by rebinding. However, the rebinding rate would be expected to decrease in the presence of identical receptors in solution. Our theoretical considerations predict that rebinding will be prevented if the following inequality is satisfied:

$$S \gg N^2 k_{on}/(16\pi^2 Da^4) \tag{23}$$

where S is the free receptor site concentration in solution, and N, k_{on}, D, and a are as defined for Eq. (22). As indicated in Eq. (23), the predicted concentration of solution receptors needed to prevent rebinding is proportional to N^2, the square of the cell surface receptor density. For RBL cells with 6×10^5 binding sites per cell, this theory predicts that to prevent DCT rebinding to cell surface IgE during dissociation requires $S \gg 2400$ nM. We showed experimentally that for $S = 200–1700$ nM, the dissociation rate of DCT from surface IgE is still substantially slower than from solution IgE where no rebinding occurs (Goldstein et al., 1989).

We have also used unlabeled IgE as competitor to analyze dissociation of (DCT)$_2$-cys from FITC-IgE in solution and on the cell surface. This competition is potentially complicated. For example, additional cross-linking can occur in the presence of unlabeled IgE, since it may bind to (DCT)$_2$-cys that is also bound to FITC-IgE. In this case the dissociation process is expected to have some dependence on the concentration of the unlabeled IgE. This type of cross-linking has been detected experimentally for IgE in solution (Posner, Erickson, Holowka, Baird, and Goldstein, submitted for publication), but not for IgE on cells (Erickson, 1988). When FITC-IgE is confined to the cell surface, additional uncertainty arises with the unlabeled IgE as competitor because its large size may prevent it from binding readily to cell-associated bivalent ligands. For bivalently bound (DCT)$_2$-cys, two ends must dissociate before this ligand can escape from the cell, and both dissociation steps must be prevented from reversing by a perfect competitor. Recent results from our laboratory suggest that unlabeled IgE (at high concentrations) does not effectively prevent re-formation of the bivalently bound state, while the monovalent ligand DCT is an effective competitor for this step (Erickson, Posner, Goldstein, Holowka, and Baird, manuscript in preparation). Equation (19) does not account for these complicating effects, and so this equation may not be strictly applicable when unlabeled IgE is used as a competitor in dissociation experiments with (DCT)$_2$-cys. We find that Eq. (19) is still useful for quantifying the binding data in an approximate manner, yielding apparent values for the parameters that are probably averages over several dynamic processes. Although the physical meaning of these apparent values may be obscured by rebinding events, they are useful in their ability to reflect different stages of cell surface receptor clustering.

Since cross-linking of IgE on the cell surface potentially stimulates cellular events that may affect the binding conditions, we can use binding measurements to detect such events. For example, we have found that for cell-bound FITC-IgE, preincubation with (DCT)$_2$-cys for progressively longer times leads to a time-dependent decrease in the dissociability of this ligand (Baird et al., 1988; Erickson, 1988). Figure 9a illustrates this result with the dissociation curves from three identical samples that were preincubated with the same concentration of (DCT)$_2$-cys for 10, 60, and 120 min before addition of excess soluble anti-DNP IgE to initiate dissociation. We consistently find that the bivalent ligands dissociate less readily with longer preincubation times, and the dissociation curves appear to get closer together at the longer times. Under the conditions of Fig. 9a, no change in dissociation

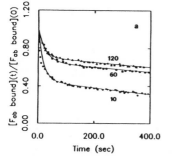 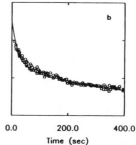

FIGURE 9. Kinetics of dissociation of $(DCT)_2$-cys (a) from cell-bound IgE after 10 min, 60 min, or 120 min preincubation time (data points fall on distinct curves as indicated); or (b) from vesicle-bound IgE after 10 min or 135 min preincubation (data points fall on the same curve). Cell samples contained 0.2 nM IgE (3.2×10^6 cells) and 0.91 nM DNP. Vesicle samples contained 5.5 nM IgE and 6.4 nM DNP. In all samples dissociation was initiated by adding excess unlabeled IgE ($>1 \mu M$) to the suspension. The normalized data are fit with Eq. (19) to yield values for the apparent fraction of cross-linked species at the initiation of dissociation ($2 [Z](0)$). For the cells these values are 0.46 (10 min), 0.66 (60 min), and 0.69 (120 min); for the vesicles the same value of 0.49 is obtained for both the 10-min and the 135-min samples. From Erickson (1988).

kinetics with preincubation time is observed for the monovalent DCT ligand in parallel samples. Experiments performed at 8°C exhibit a similar decrease in the rate of $(DCT)_2$-cys dissociation with preincubation as shown in Fig. 9a, although the time course for this change is somewhat slower.

The time-dependent loss in dissociability for $(DCT)_2$-cys is not observed in plasma membrane vesicle preparations derived from RBL cells (Fig. 9b). These preparations are predominantly right-side-out, sealed vesicles that lack an intact cytoskeleton but retain the IgE receptor as well as other plasma membrane components (Holowka and Baird, 1983). The observed differences between cells and vesicles in the time-dependent dissociability of $(DCT)_2$-cys cannot be explained by possible internalization occurring with the cells since the same result is obtained at low temperatures and in the presence of metabolic inhibitors. Cytoskeletal involvement is a possible explanation. Previous studies have shown that cross-linked receptors on cells but not on vesicles can become immobilized, associated with the cytoskeleton and sometimes gathered together into larger clusters (patches) on the cell surface (Baird *et al.*, 1988). Although patching has been detected with $(DCT)_2$-cys only after several hours, immobilization and some smaller amount of clustering induced by the cell could occur at early times, and the corresponding change in receptor density is a plausible explanation for the time-dependent change in the dissociation properties.

The time-dependent loss of dissociability for $(DCT)_2$-cys from cells can be mimicked by reagents that are known to cross-link IgE-receptor complexes to each other. Figure 10 shows the effect of two such treatments and compares the results to a control sample containing only $(DCT)_2$-cys. Concanavalin A (Con A) is a tetravalent plant lectin that binds and cross-links glycoproteins and glycolipids on cell membranes including glycosylated regions of receptor-bound IgE (Fewtrell, *et al.*, 1979). Con A was added simultaneously with $(DCT)_2$-cys, and this results in a marked change in the dissociation rate upon the subsequent addition of unlabeled IgE. There is an even greater effect on $(DCT)_2$-cys dissociation by polyclonal rabbit anti-IgE. This reagent by itself provides a strong stimulus for secretion at this concentration through extensive cross-linking of receptor-bound IgE (Estes *et al.*, 1987). The large effects on $(DCT)_2$-cys dissociation from IgE on the cell surface by Con A and anti-IgE are likely due to the enhanced probability of IgE being cross-linked and clustered. Enhanced cross-linking is consistent with the observed increased initial rate of fluorescence quenching of FITC-IgE by $(DCT)_2$-cys in the presence of Con A and anti-IgE (Erickson, 1988). Parallel samples with DCT showed no detectable difference in association or dissociation kinetics in the presence of Con A or anti-IgE, indicating that the observed effects are not steric in nature.

6. CONCLUDING REMARKS

This chapter describes fluorescence methodology for investigating ligand binding to and cross-linking of receptors in solution and on cell surfaces. The strength of the method rests on a variety of factors including: (1) a well-defined experimental system which allows a large degree of experimen-

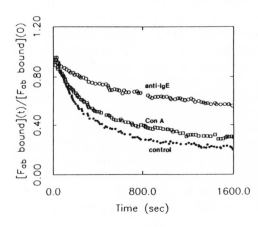

FIGURE 10. Kinetics of dissociation of $(DCT)_2$-cys from cell-bound IgE after 15 min preincubation in the presence of anti-IgE (O), concanavalin A (□), or no additional reagent (●). From Erickson (1988).

tal control; (2) use of fluorescence spectroscopy which allows continuous monitoring of ligand binding; (3) a high signal-to-noise ratio which allows low concentrations of labeled species to be monitored. The highly sensitive fluorescence method allows kinetic binding studies of high-affinity interactions (such as those between DCT or $(DCT)_2$-cys and anti-DNP IgE) that are characterized by forward rate constants $> 10^7 \, M^{-1} \, sec^{-1}$.

The method directly reveals some qualitative features and trends in the binding properties that are model independent. For example, in the equilibrium studies the dose-dependence for $(DCT)_2$-cys binding to cell surfaces clearly has a steeper initial slope than that for DCT binding indicating that cross-linking occurs with the bivalent ligand (Fig. 5). The data can be fit to models of ranging complexity to provide interpretation of the binding features. In principle, any type of defined complexity can be included in a theoretical binding model. However, the value of applying a complicated model is limited by the range and quality of the data that can be achieved experimentally. The minimal models described in this chapter can be used to fit the binding data and yield equilibrium and kinetic constants. For the cases of monovalent ligand binding described in this chapter, the simple models are probably good representations. For example, in the kinetic studies of monovalent ligand binding to receptors on the cell surface (Fig. 7), the fitted curve yields density-dependent rate constants that agree well with independently determined values. The extrapolated k_{on}(per receptor) at very low receptor density is the same as k_{on} measured directly for IgE in solution; the extrapolated k_{on}(per cell) at high receptor density is consistent with the diffusion-limited value predicted for the ligands and cells involved (Section 5.3.1). In applications such as bivalent ligands binding to and cross-linking receptors on the cell surface, there is probably greater complexity in the system than is provided by the minimal model, particularly if the cell changes in response to these binding events. The parameters obtained from the theoretical curves in these cases reflect some kind of average. However, these averaged parameters are still useful for quantifying changes that are occurring and thereby altering the binding conditions. For example, the time dependence of the rate of dissociation of $(DCT)_2$-cys from cell surface IgE (Fig. 9a) is characterized by the changes in the values obtained for $2[Z(0)]$ from application of Eq. (19). The observed slowing of $(DCT)_2$-cys dissociation suggests that some cellular change occurs after the bivalent ligands bind and cross-link IgE receptors on the cell surface.

With the experimental system of DNP ligands and anti-DNP IgE, we have demonstrated how detailed binding measurements based on fluorescence quenching can provide insight to essential features of ligand interactions with receptors on cell surfaces and subsequent signal transduction events. Greater elucidation of the mechanisms involved in these processes will come from refinements in both the experimental and the theoretical aspects. For example, a more detailed data set can be obtained by continuous data acquisition facilitated by a computer interface (Goldstein *et al.,* 1989). Simultaneously, extended models can be developed to explore the relationship of the derived binding parameters to events following receptor cross-linking. In this regard it will be interesting to investigate the possibility that the observed reduction in dissociability of bivalent ligands from cell surface receptor is related to cytoskeletal association of the cross-linked receptors.

ACKNOWLEDGMENTS. This work was supported by National Institutes of Health Grants GM35556, AI18306, and AI22449, and by the United States Department of Energy.

REFERENCES

Baird, B., and Holowka, D., 1988, in *Spectroscopic Membrane Probes* (L. Loew, ed.), CRC Press, Boca Raton, pp. 93–116.

Baird, B., Erickson, J., Goldstein, B., Kane, P., Menon, A. K., Robertson, D., and Holowka, D., 1988, in *Theoretical Immunology* (A. Perelson, ed.), Addison–Wesley, Reading, Mass., pp. 41–59.

Barsumian, E. L., Isersky, C., Petrino, M. G., and Siraganian, R. P., 1981, *Eur. J. Immunol.* **11**:317–323.

Berg, H. C., and Purcell, E. M., 1977, *Biophys. J.* **20**:193–219.

Carson, D., and Metzger, H., 1974, *Immunochemistry* **11**:355–359.

Dembo, M., and Goldstein, B., 1978a, *Immunochemistry* **15**:307–313.

Dembo, M., and Goldstein, B., 1978b, *J. Immunol.* **121**:345–353.

Dembo, M., Goldstein, B., Sobotka, A. K., and Lichtenstein, L. M., 1979, *J. Immunol.* **123**:1864–1871.

Eisen, H., and McGuigan, J., 1968, in *Methods in Immunology and Immunochemistry* (W. C. Williams and M. Chase, eds.), Academic Press, New York.

Erickson, J. W., 1988, Equilibrium and kinetic studies of a model ligand–receptor system: Monovalent and bivalent ligand interactions with immunoglobulin E, Ph.D. thesis, Cornell University.

Erickson, J., Kane, P., Goldstein, B., Holowka, D., and Baird, B., 1986, *Mol. Immunol.* **23**:769–780.

Erickson, J., Goldstein, B., Holowka, D., and Baird, B., 1987, *Biophys. J.* **52**:657–662.

Estes, K., Monfalcone, L., Hammes, S., Holowka, D., and Baird, B., 1987, *J. Cell Biol.* **105**:747–755.

Fewtrell, C., 1985, in *Calcium in Biological Systems* (G. Weiss, J. Putney, and R. Rubin, eds.), Plenum Press, New York, pp. 129–136.

Fewtrell, C., Kessler, A., and Metzger, H., 1979, *Adv. Inflam. Res.* **1**:205–221.

Flory, P. J., 1953, *Principles of Polymer Chemistry,* Cornell University Press, Ithaca, N.Y.

Goldstein, B., Posner, R. G., Torney, D. C., Erickson, J., Holowka, D., and Baird, B., 1989, *Biophys. J.* **56**:955–966.

Haugland, R. P., 1989, *Molecular Probes Handbook of Fluorescent Probes and Research Chemicals,* Molecular Probes, Eugene, Ore.

Holowka, D., and Baird, B., 1983, *Biochemistry* **22**:3466–3474.

Ishizaka, T., Hirata, F., Ishizaka, K., and Axelrod, J., 1981, in *Biochemistry of the Acute Allergic Reactions* (E. Becker, A. Simon, and K. Austen, eds.), Liss, New York, pp. 213–227.

Isersky, C., Rivera, J., Mims, S., and Triche, T., 1979, *J. Immunol.* **122**:1926–1936.

Kane, P., Erickson, J., Fewtrell, C., Baird, B., and Holowka, D., 1986, *Mol. Immunol.* **23**:783–790.

Kane, P., Holowka, D., and Baird, B., 1988, *J. Cell Biol.* **107**:969–980.

Liu, F. T., Bohn, J. W., Ferry, E. L., Yamamoto, H., Molinaro, C. A., Sherman, L. A., Klinman, N. R., and Katz, D., 1980, *J. Immunol.* **124**:2728–2735.

Menon, A. K., Holowka, D., and Baird, B., 1984, *J. Cell Biol.* **9**:577–583.

Menon, A. K., Holowka, D., Webb, W. W., and Baird, B., 1986a, *J. Cell Biol.* **102**:534–540.

Menon, A. K., Holowka, D., Webb, W. W., and Baird, B., 1986b, *J. Cell Biol.* **102**:541–550.

Metzger, H., Alcaraz, G., Hohman, R., Kinet, J.-P., Pribluda, V., and Quarto, R., 1986, *Annu. Rev. Immunol.* **4**:419–470.

Pecht, I., and Lancet, D., 1977, in *Chemical Relaxation in Molecular Biology* (I. Pecht and R. Rigler, eds.), Springer-Verlag, Berlin, pp. 306–338.

Perelson, A., and DeLisi, C., 1980, *Math. Biosci.* **48**:71–110.

Robertson, D., Holowka, D., and Baird, B., 1986, *J. Immunol.* **136**:4565–4572.

Schweitzer-Stenner, R., Light, A., Luscher, I., and Pecht, I., 1987, *Biochemistry* **26**:3602–3612.

Chapter 7

Fluorescence Energy Transfer in Membrane Biochemistry

T. Gregory Dewey

1. INTRODUCTION

In the two decades since its first application to biological systems (Stryer and Haugland, 1967), fluorescence energy transfer has become a standard technique for measuring distances in biological systems. Early studies usually measured distances between a single donor and a single acceptor each at a specific location. The extension to transfer between multiple donors and multiple acceptors has proven quite useful for the study of multienzyme complexes (Hahn and Hammes, 1978; Angelides and Hammes, 1979). Most of the applications to membrane biochemistry also represent a situation in which multiple donors and multiple acceptors are present. These applications have a long history as well, starting with the initial study of energy transfer from chlorophyll in monomeric films (Tweet *et al.*, 1964). This early work relied heavily on Foerster's derivation of energy transfer between multiple acceptors and multiple donors in three dimensions (Foerster, 1949) and derived the appropriate expression for two dimensions. Unfortunately, these results went unnoticed and identical expressions have been derived in several subsequent works. The first studies on lipid bilayer systems were aimed at determining the depth of a chromophore in the membrane (Shaklai *et al.*, 1977) or at determining the surface density of an acceptor (Fung and Stryer,

T. GREGORY DEWEY • Department of Chemistry, University of Denver, Denver, Colorado 80208.

1978). These have been the two observables of major interest in membrane systems. Energy transfer has subsequently been used to monitor a variety of membrane processes and interactions. A representative example of the diversity of applications is given in Table I. Additional references may be found in recent reviews (Hammes, 1981; Blumberg, 1985; Tron *et al.*, 1987).

This chapter concentrates on recent developments and applications of fluorescence energy transfer to membrane systems. First, the theoretical background and analysis of energy transfer in two-dimensional systems are dealt with. The two-dimensional theory has presented special problems and consequently developed a literature of its own. Recently, several exciting new approaches to the problem have appeared (Ediger and Fayer, 1983; Klafter and Blumen, 1984) and there is hope for a more facile analysis than those employed in the past. The second section deals with the development of a new technique to kinetically resolve changes in fluorescence energy transfer due to conformational processes in membrane-bound proteins. Using phase-modulation of fluorescence energy transfer, specific structural changes have been observed in bacteriorhodopsin as it proceeds through its proton-pumping photocycle (Hasselbacher *et al.*, 1986; Hasselbacher and Dewey, 1986). This technique provides a rare combination of a specific structural measurement with good kinetic resolution. The next section considers the case of fluorescence energy transfer from excimer states. There is an increasing need in membrane biochemistry to develop techniques for monitoring the formation of ternary complexes, e.g., receptor–G protein–

TABLE I. Applications of Fluorescence Energy Transfer to Membrane Systems

Model systems
  Alloxazines in liposomes (Aso *et al.*, 1980)
  Anionic surfactants (Kano *et al.*, 1981)
  Anthracene in bilayers (Shimomura *et al.*, 1982)
  Perfluorosulfonate membranes (Nagata *et al.*, 1983)
  DPH in liposomes (Davenport *et al.*, 1985)
  Carbazole in bilayers (Kunitake *et al.*, 1985)
  Naphthalene in bilayers (Nakashima *et al.*, 1985)
  Rhodamine adsorbed on vesicles (Tamai *et al.*, 1987)
  Porphyrins adsorbed on vesicles (Takami and Mataga, 1987)
  Cyanine dyes in bilayers (Nakashima *et al.*, 1987)
  Surface density measurements in vesicles (Fung and Stryer, 1978)
Membrane structure and processes
  Membrane thickness (Peters, 1971)
  Valinomycin-induced changes (Bessette and Seufert, 1978)
  Lymphocyte membranes (Mani *et al.*, 1975)
  Nerve membranes (Tasaki *et al.*, 1976)
  Cell fusion (Keller *et al.*, 1977)

TABLE I. *(Continued)*

Membrane structure and processes (*continued*)
    Cholesterol in bilayers (Rogers *et al.*, 1979)
    Vesicle fusion (Gibson and Loew, 1979; Vanderwerf and Ullman, 1980; Struck *et al.*, 1981)
    Sialoglycoprotein and spectrin (Nakajima *et al.*, 1979)
    Phase behavior in rod outer segments (Sklar *et al.*, 1979)
    Intervesicle exchange (Kano *et al.*, 1980)
    Membrane freezing (Kirchanski and Branton, 1980; MacDonald and MacDonald, 1983)
    Myoblast fusion (Herman and Fernandez, 1982)
    Chromaffin granules (Morris *et al.*, 1982; Morris and Bradley, 1984)
    Plasma membrane vesicles (Holowka and Baird, 1983)
    Complement polymerization (Sims, 1984)
    Virus–membrane fusion (Chejanovsky *et al.*, 1984)
    Tissue dehydration (Smith *et al.*, 1986; Womersley *et al.*, 1986)
    Nuclear transport (Arvinte *et al.*, 1987)
    Energy transfer microscopy (Uster and Pagano, 1986)
    Microsomal vesicles (Comerford and Dawson, 1988)
Membrane-to-protein distances
    Gramicidin (Veatch and Stryer, 1977; Haigh *et al.*, 1979)
    Peripheral protein association to erythrocytes (Nakajima *et al.*, 1979)
    Tryptophan in cytochrome b_5 (Fleming *et al.*, 1979; Kleinfeld and Lukacovic, 1985)
    Hemoglobin (Shaklai *et al.*, 1977; Eisinger and Flores, 1982)
    Lactalbumin (Dangreau *et al.*, 1982)
    Bacteriorhodopsin (Oesterhelt *et al.*, 1981; Rehorek *et al.*, 1983; Kouyama *et al.*, 1983;
        Chatelier *et al.*, 1984; Hasselbacher *et al.*, 1986; Hasselbacher and Dewey, 1986)
    Cytochrome P450 (Omata *et al.*, 1987)
    Calcium ATPase (Gutierrez-Merino *et al.*, 1987; White and Dewey, 1987)
    Tryptophan in membrane proteins (Trung *et al.*, 1983)
    Coagulation factor V_a (Isaacs *et al.*, 1986)
    Sodium–glucose cotransporter (Peerce and Wright, 1986)
    Band 3 (Dissing *et al.*, 1979)
    ADP/ATP carrier (Graue and Klingenberg, 1979)
Protein–protein interactions
    Calcium ATPAse (Vanderkooi *et al.*, 1977; Fagan and Dewey, 1986; Highsmith and Cohen,
        1987)
    Complement proteins C5b–8 (Chen *et al.*, 1985)
    Bacteriorhodopsin (Hasselbacher *et al.*, 1984; Kouyama *et al.*, 1983; Kometani *et al.*, 1987)
    Lectin receptors (Chan *et al.*, 1979)
    Gram-negative membrane proteins (Hyono *et al.*, 1985)
    Melittin (Talbot *et al.*, 1987)
Theoretical
    Analytic theory in 2-D (Tweet *et al.*, 1964; Shaklai *et al.*, 1977; Kampmann, 1977; Estep
        and Thompson, 1979; Koppel *et al.*, 1979; Wolber and Hudson, 1979)
    Approximations and different geometries (Dewey and Hammes, 1980)
    Monte Carlo calculations (Snyder and Freire, 1982; Wolber and Hudson, 1979)
    Fractals (Tamai *et al.*, 1986; Klafter and Blumen, 1984; Dewey and Datta, 1989)
    Aggregates (Hasselbacher *et al.*, 1984)
    Tryptophan imaging (Kleinfeld, 1985)
    Finite volume systems (Ediger and Fayer, 1983, 1984)

enzyme complexes. The formation of a binary complex can be monitored by labeling the components with excimer-forming probes attached via flexible chains. Proximity of the two components can then be monitored by observing excimer fluorescence. The proximity of an additional, third component can, in turn, be observed by energy transfer from the excimer state to an acceptor labeling the third component. In this section the theoretical and experimental bases for such a technique are established. In the fifth section, the analysis of fluorescence energy transfer from exciton states is considered. There have been a number of applications which have involved energy transfer from multiple donors in close proximity to each other. These situations occur for energy transfer from antennae pigments to reaction centers. They also may occur when observing energy transfer from multiple tryptophans to an acceptor. In these examples the donor may be close enough either to have significant self-transfer or to form exciton states as a result of dipolar interactions. As a practical consideration it is important to treat exciton effects in the analysis of energy transfer. In the final section, prospectives on future applications of fluorescence energy transfer to biochemistry and cell biology are given.

2. DATA ANALYSIS AND INTERPRETATION

The starting point for theoretical treatments of fluorescence energy transfer between multiple donors and multiple acceptors is the following rate equation for fluorescence decay (see Shaklai et al., 1977):

$$\frac{dp_D(t)}{dt} = -\{k_e + k_i + \sum_{j=1}^{N} k_j(r)p_A(t)\} \tag{1}$$

where $p_D(t)$ is the probability that a donor molecule excited at time $t = 0$ is still excited at time t, k_e and k_i are the specific rate constants for emission and internal quenching, respectively, k_j is the rate constant for energy transfer to an acceptor A located at a distance r, and p_A is the probability of finding the acceptor at distance r at time t. The summation runs over the N acceptor molecules. From Foerster's theory the rate of energy transfer is given by (Foerster, 1965):

$$k(r) = (R_o/r)^6(1/\tau_o) \tag{2}$$

where $1/\tau_o = (k_e + k_i)$ and R_o is the characteristic Foerster distance. At this point one must solve Eq. (1) and perform an ensemble average. The proper ensemble average as first presented by Foerster (1949) contains subtle points that have been ignored in some erroneous theoretical treatments which have appeared in the literature. In this average it is assumed that there is a uniformly distributed population of acceptors. However, at any given time each donor is surrounded by a slightly different acceptor population than any other donor. Thus, the averaging process must occur over both donor and acceptor distributions. When the proper average is taken for uniform distributions of donors and acceptors in two dimensions, the following expression is obtained (Shaklai *et al.*, 1977):

$$p_D(t) = e^{-t/\tau_0}e^{-\sigma L(t)} \tag{3}$$

where

$$L(t) = \int_{R_e}^{\infty} (1 - e^{-(t/\tau_0)(R_0/r)^6})2\pi r^{d-1} \, dr$$

The parameter σ is the surface density of acceptors and R_e is the distance of closest approach of a donor to an acceptor. For a continuous two-dimensional distribution of acceptors, $d = 2$ (Fung and Stryer, 1978). For a random distribution on a fractal structure, d is equal to the fractal dimension (Klafter and Blumen, 1984). Wolber and Hudson (1979) presented the equation for $d = 2$ using an expression which involved incomplete gamma functions. It can be shown that these two forms are equivalent by performing an asymptotic expansion of Eq. (3) followed by a term-by-term integration. Unfortunately, these forms of the equation are mathematically complicated expressions which are not easily adapted to data fitting routines. The fluorescence decay that is observed in these cases is nonexponential and is of the form (see Tamai *et al.*, 1986):

$$p_D(t) = e^{-t/\tau_0}e^{-\gamma(t/\tau_0)^\beta} \tag{4}$$

where γ and β are constants. Most studies do not actually fit the time course of the fluorescence decay; rather, the experimentally determined parameter is the efficiency of energy transfer, E, which is defined as

$$E = 1 - \tau_{DA}/\tau_D = 1 - Q_{DA}/Q_D \tag{5}$$

where τ_{DA} and τ_D are the average relaxation times of the donor in the presence and absence of acceptor, respectively. Q_{DA} and Q_D are the quantum yields of the donors in the presence and absence of acceptor, respectively. The quantum yield ratio will be equal to the ratio of the steady-state fluorescence intensities. Thus, the efficiency may be determined from either transient or steady-state data. In simple cases of energy transfer between a single donor–acceptor pair, the measurement of fluorescence lifetimes has been favored over steady-state measurements. Heterogeneous populations and quenching may be more easily identified with the time-resolved decay. In the case of multiple donors and acceptors the decay is already a complicated function and it is difficult to distinguish this behavior from that of a sum of exponentials. In most cases the average relaxation time, τ_{ave}, is obtained by first fitting the decay curve to a sum of exponentials with amplitudes A_i and relaxation times τ_i. The average relaxation time is then defined by:

$$\tau_{ave} = \sum A_i \tau_i / \sum A_i \tag{6}$$

The efficiency defined in the above manner can then be related to the following theoretical expression for energy transfer in the two-dimensional case:

$$E = 1 - (1/\tau_o) \int_0^\infty p_D(t)\, dt \tag{7}$$

In most applications, Eq. (7) is solved by numerical integration (Fung and Stryer, 1978; Wolber and Hudson, 1979). A tabulation of these numerical results is given in Wolber and Hudson's paper and provides a convenient means to analyze most data. It is also possible to avoid these analytic expressions and analyze the data using Monte Carlo calculations (Wolber and Hudson, 1979; Snyder and Freire, 1982).

In applying Eq. (7), it is important to be aware of the assumptions of the model and to be careful to establish that the experimental conditions match these assumptions. These assumptions are detailed below and their implications for the interpretation of results are discussed.

1. Donors and acceptors are considered to be uniformly distributed on two planar two-dimensional surfaces separated by a distance R_e, the distance of closest approach. The distance R_e may be taken to be zero. Accurate models exist for this special case (Estep and Thompson, 1979; Wolber and Hudson, 1979) and they will not be considered here.

2. The characteristic Foerster distance R_o is independent of time and distance and all fluorescence energy transfer occurs via a dipole–dipole interaction mechanism.

3. Donors and acceptors are in low concentrations so that excluded volume effects need not be considered. Also, donors must be in low enough concentration that self-transfer does not occur.

Assumption 1 contains features that at first seem unrealistic when considering energy transfer from lipids to an acceptor embedded in a membrane protein. The distance of closest approach in this situation is the shortest distance from the lipid-accessible periphery of the protein to the acceptor location. It is not the distance between the actual planes in which the donor and acceptor are located. Thus, the volume excluded by the protein contributes to the distance of closest approach. However, this specific model does not account for the fact that acceptors are excluded from regions in their own plane by the presence of the protein. A theoretical model by Koppel *et al.* (1979) incorporated this feature into it. Analysis of their experimental results for cytochrome b_5 (Fleming *et al.*, 1979) indicated that the calculated distance of closest approach was affected very little by large changes in estimates of the membrane surface area excluded by the protein. There was only a 10% change in R_e when going from no excluded area to a 1250 Å^2 area. These results would indicate that the distance of closest approach is the most sensitive parameter while the size and shape of the protein have surprisingly little influence on the fluorescence quenching.

A second potential problem with the first assumption is that of the uniform distributions of donors and acceptors. Many membrane-bound proteins aggregate either forming specific oligomeric complexes or forming a separate protein domain or phase. This aggregation can have a dramatic effect on fluorescence energy transfer from lipids to protein as has been demonstrated for the case of bacteriorhodopsin reconstituted into liposomes (Hasselbacher *et al.*, 1984). A means of testing for protein–protein interactions is to locate the donor on the protein and the acceptor in the lipid and then demonstrate that energy transfer is independent of donor concentration. Even in situations in which monomeric donors and acceptors are assured, there may be reasons to question the assumption of uniform distributions. In recent work on energy transfer between donors and acceptors adsorbed to a vesicle surface, it was shown that the time-resolved decay could not be fit by assuming a uniform distribution (Tamai *et al.*, 1986). These curves could be analyzed with a model that assumes a random distribution on a fractal structure (Klafter and Blumen, 1984). The decay curves were analyzed using Eq. (4) where γ and β were treated as adjustable parameters. The fractal dimension, d, is determined by $\beta = d/s$ where s is the order of the multipolar interaction. For Foerster transfer s is equal to 6. In these experiments the fractal dimension was found to be independent of specific donor–acceptor pairs and was equal to 1.3 (as opposed to 2 for a continuous uniform distribution). The fractal dimension for a self-avoiding random walk is also 1.3. The structural interpretation of these results is not clear, although they do suggest long-range correlation in the adsorption of dye molecules onto vesicles. Similar analyses have yet to be done on a protein–lipid system. Most work to date can be satisfactorily analyzed with the continuum model. However, this may be a result of not extending the experimental parameters over a wide enough range. The largest deviations from fractal and continuum behavior occur in the region of greater than 90% quenching (Tamai *et al.*, 1986).

The second assumption is the least troublesome. The characteristic Foerster difference should not depend upon the distance between donor and acceptor. This will certainly be true as long as the distance of closest approach is greater than the size of the chromophores. Dexter (1953) provided a theoretical description which accounts for the breakdown of the electrostatic moment expansion at short distances. This work suggests that Foerster behavior should dominate even at distances less than the radii of the donors and acceptors provided a strongly allowed transition occurs for the donor. In most applications, this presents no problem. However, these conditions may not hold when considering the technique of diffusion-enhanced fluorescence energy transfer (Stryer et al., 1982). In these experiments, donors with exceptionally long fluorescence lifetimes are used so that acceptor may translationally diffuse close to the donor. However, long fluorescence lifetimes are generally an indication of a forbidden transition. In these cases the dipolar mechanism contributes weakly to the decay and may be comparable in strength to quadrupolar and magnetic dipolar mechanisms. In addition to causing problems because of deviations from the Foerster transfer mechanism, it would be extremely difficult to obtain an accurate R_0 for such cases because the fluorescence spectrum used to calculate the overlap integral does not result from a dipolar decay process. Another source of variation in R_0 may result from donors in different spatial locations in the membrane with different quantum yields. This would occur for lipid donors in systems where separate phases coexist. However, a large change in the donor's quantum yield would have to occur to profoundly influence R_0. If all other parameters stay constant, a doubling of the quantum yield increases the characteristic Foerster distance by only 12%. Likewise, large variations in the dipolar orientation factor, κ^2, must occur for a significant change in R_0. While κ^2 may show some time dependence, it probably does not change enough on the fluorescence time scale to seriously affect the calculation. This parameter has been discussed extensively in the literature and several analyses for membrane systems have been published (Davenport et al., 1985; Koppel et al., 1979; Hasselbacher and Dewey, 1986).

The final assumption is that the donors and acceptors are in low concentrations. It is particularly important that the donor be in low enough concentration that self-transfer does not occur. The efficiency of energy transfer should be independent of donor concentration so a check against self-transfer is to establish a concentration regime in which the efficiency is constant. Fluorescence depolarization decays may also be used as a monitor of self-transfer (see Ediger and Fayer, 1984). Working with rhodamine in a micellar system, it was shown that self-transfer persisted at surface densities as low as 10^{-5} molecule/Å^2. However, Fung and Stryer (1978) used surface densities of donor as high as 10^{-3} molecule/Å^2 and established no dependence of transfer efficiency on donor concentration. Undoubtedly self-transfer will depend markedly on the R_0 for self-transfer and the geometry of the system. The most difficult situation arises when observing transfer from tryptophans within a protein. In this case there is no way (short of site-directed mutagenesis) of "diluting" the tryptophan concentration. Even though the R_0 for self-transfer may be small, significant transfer can occur because of the proximity of the donors. This situation is discussed again in Section 5.

While the theoretical models of fluorescence energy transfer in membranes are admittedly simple, they nevertheless are applicable to a wide range of experimental systems. Frequently, a system of more complex geometry must be considered. Two such cases are shown in Fig. 1. In the first case, energy transfer was measured from the chloroplast H^+-ATPase which had been coated with fluorescent antibodies to the membrane surface (Baird *et al.*, 1979). This situation was modeled as two separated spheres, one coated with donors and the other with acceptors. In the second situation, energy transfer was observed from lipid donors to an acceptor located on a membrane-bound protein which aggregates (Hasselbacher *et al.*, 1984; Fagan and Dewey, 1986). This was modeled as energy transfer from donors on a planar surface to acceptors uniformly distributed within a circular disk. In such

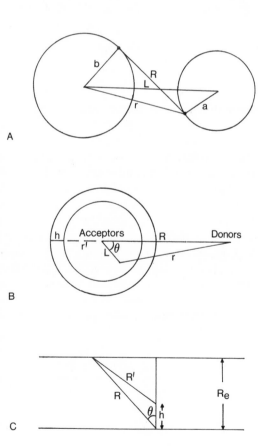

FIGURE 1. Diagram showing different geometries for membrane energy transfer. (A) Separated spheres of radii *a* and *b* whose centers are separated by a distance *L*. *R* is the distance between a given donor–acceptor pair. (B) Disk containing acceptors with radius *r'* and annulus *h*. Area outside annulus contain a uniform distribution of donors. *L* is distance from center to a given acceptor, *R* is distance from center to a given acceptor, and *r* is distance between a given donor–acceptor pair. (C) Distance change due to conformational process in a membrane system. R_e is distance of closest approach to acceptor in original conformation, *h* is change in distance as a result of the conformational change, *R* and *R'* are distances between a given donor and the acceptor in the original and final conformational state, respectively.

cases it is not always easy to apply analogous theoretical approaches as in the simple planar model. Consequently, an approximation method was developed to treat these more complicated situations (Dewey and Hammes, 1980). This method allows one to derive simple mathematical expressions for each approximant and it has the advantage that the error in the approximation can be determined directly.

A convenient starting point for the derivations involving more complicated geometries is the steady-state expression for the quenching ratio:

$$Q_{DA}/Q_D = (1/N_D) \sum_i [1 + \sum_j (R_o/R_{ij})^6]^{-1} \qquad (8)$$

where R_{ij} is the distance from the ith donor to the jth acceptor and N_D is the number of donors. Equation (8) is expanded in a series and the summations are converted into integrals over the specific areas for which donors and acceptor are uniformly distributed. Using the method of Gordon (1968), the quenching ratio may be expressed as a continued fraction:

$$\frac{Q_{DA}}{Q_D} = \cfrac{\alpha_1}{1 + \cfrac{\alpha_2}{1 + \cfrac{\alpha_3}{1 + \cdots}}} \qquad (9)$$

where the first three coefficients are:

$$\alpha_1 = 1; \qquad \alpha_2 = \mu_1; \qquad \alpha_3 = (\mu_2 - \mu_1^2)/\mu_1$$

In the two specific cases shown in Fig. 1, μ_i represents the ith moment of the expansion of Eq. (8) and is given by:

$$\mu(n) = (1/S_D) \int_{S_D} \left\{ N_A/S_A \int_{S_A} (R_o/r)^6 \, dS_A \right\}^n dS_D \qquad (10)$$

where S_D and S_A are the surface area covered by the donors and acceptors, respectively, and r is the distance from a point in one area to a point in the second one (see Fig. 1). For the two cases mentioned, the first two moments are:

Transfer between separated spheres of radii a and b and separation, L:

$$\mu_1 = N_A R_o^6$$

$$\times \frac{L^6 - L^4(a^2 + b^2) + L^2[8a^2b^2/3 - (b^2 - a^2)^2] + (b^2 - a^2)^2(a^2 + b^2)}{[L^4 - 2L^2(a^2 + b^2) + (b^2 - a^2)^2]^3}$$

$$(11)$$

$$\mu_2 = \{N_A^2 R_o^{12}/4aL\} \frac{(4/7)b^4(y^7 - x^7)}{y^7x^7} + \frac{(2/3)b^2(y^6 - x^6)}{y^6x^6}$$

$$+ \frac{(1/5)(y^5 - x^5)}{y^5x^5}$$

where $y = L^2 - b^2 + a^2 + 2aL$ and $x = L^2 - b^2 + a^2 - 2aL.$

Transfer between a plane and a disk of radius r' with an annulus of width, h:

$$\mu_1 = \{\pi^2 N_A R_o^6/2S_D S_A h^3\}[(r' + h)^2 r'^2/(2r' + h)^3] \tag{12}$$

$$\mu_2 = \{\pi^3 N_A R_o^{12} r'^4/4S_D S_A h^7(2r' + h)^7\}$$

$$[9r'^4/7 + 4hr^3 + 26r'^2h^2/5 + 16r'h^3/5 + 4h^4/5]$$

Using these moments the first two continued fractions, A_1 and A_2, may be calculated by setting all α's of higher order equal to zero. The best estimate for Q_{DA}/Q_D is then $(1/2)(A_1 + A_2)$ and the error in the estimate is $(1/2)|A_1 - A_2|$. The error in using only the first two approximants is very dependent upon the specific geometric parameters. However, in most realistic cases they are 20% or less.

Using the separated sphere model, the structure of the chloroplast ATP-ase was investigated in a reconstituted system (Baird *et al.*, 1979). Antibodies raised to CF_1 (the soluble head portion of the ATPase) or to the subunits α, β, γ, and ϵ were fluorescent labeled. Monovalent fragments of these antibodies were then used as energy transfer donors. The acceptors were dye molecules incorporated into the lipid membrane. The anti-CF_1 donors coated the spherical portion of the ATPase extending from the membrane. The distance measured from the vesicle surface to the center of this spherical region was 35 Å. This was consistent with the electron micrographs of F_1–F_0 ATPases which show the familiar knoblike structure. Energy transfer from antibodies to specific subunits gave comparable results.

The disk model was used to investigate membrane protein aggregation for reconstituted systems of bacteriorhodopsin (Hasselbacher *et al.*, 1984) and the calcium ATPase (Fagan and Dewey, 1986). Bacteriorhodopsin reconstituted into phospholipid vesicles undergoes a clear-cut transition from an aggregated state to a monomeric one. The aggregated state exists when the temperature is below the lipid phase transition temperature (Heyn *et al.*, 1981a,b). When the temperature is within a few degrees of the phase transition, the aggregates "melt" out into monomers. Energy transfer was measured from indocarbocyanine lipid donors to the retinal of the bacteriorhodopsin. Using the above analysis it was possible to estimate an average radius of the protein aggregate. This radius was temperature and lipid dependent and varied from 50 to 200 Å. These experiments were also performed as a function of the surface density of protein. Surprisingly, the aggregate radius did not increase with increasing protein-to-lipid content. The data for the entire surface density dependence could be fit by assuming a single patch size for the bacteriorhodopsin. This suggested that as the protein concentration increased, the patches did not grow in size but rather increased in number. Studies performed on the sarcoplasmic reticulum calcium ATPase were aimed at elucidating the size of the functional unit. Assessing the functional significance of protein aggregates in the membrane is a common and difficult problem in protein biochemistry. It is often very difficult to distinguish by electron microscopy a specific oligomeric structure from packing due to phase separation. Energy transfer experiments performed on the calcium ATPase were analyzed with the disk model and were again studied as a function of temperature. Unlike bacteriorhodopsin, the calcium ATPase does not undergo dramatic changes in aggregation. It was not possible to demonstrate the existence of a monomeric state. However, the energy transfer results were consistent with a tetrameric structure at low temperature breaking up to a smaller structure at high temperature.

As a final topic in this section, the work of Fayer and co-workers on finite volume systems is considered. Extensive theoretical and experimental investigations have been pursued on the effects of finite volume on self-transfer and trapping of excitation energy (see Ediger and Fayer, 1984). Initially, a diagrammatic technique was developed to solve the problem of excitation transport among donors in infinite solutions (Grochanour et al., 1979). More recently, this technique was extended to consider the case of donors and acceptors at any concentration (Miller et al., 1983). This represented a more general solution to the problem solved by Foerster (1949) where donor–donor interactions were ignored. Most recently, the theory for finite volume systems has been developed. Constraining the donors to a finite volume adds greatly to the complexity of the problem. As a result of "edge" effects, the donors are not in identical physical situations. Therefore, an additional average must be performed. A density expansion is performed and Padé approximants to a Green function solution to the rate equation are obtained (Ediger and Fayer, 1983). These approximants have been shown to be quite accurate. Interestingly, this Green function is experimentally accessible by observing the fluorescence anisotropy decay rather than the total fluorescence decay. Using well-defined micellar and polymeric systems, it was possible to analyze these data to determine the size of the system. These early results are very encouraging and give added support to the theoretical development. While fluorescence energy transfer has been measured in biological systems using fluorescence anisotropy (see Weber and Daniel, 1966;

Highsmith and Cohen, 1987), there have been few applications to date of the "finite volume" theory.

3. MONITORING CONFORMATIONAL CHANGES BY KINETIC RESOLUTION OF FLUORESCENCE ENERGY TRANSFER

In his 1978 review article, Stryer anticipated the application of fluorescence energy transfer to the study of conformational transitions:

> Transient intermediates having lifetimes of milliseconds, and sometimes even of microseconds, are within the scope of the technique. It should be feasible to detect changes in distance of a few angstroms accompanying dynamic processes such as enzymatic catalysis, muscle contraction, signal transduction, and active transport. [Stryer, 1978]

Since that time there have been limited applications of energy transfer to conformational dynamics. In this section a theoretical analysis is presented for kinetically resolved energy transfer changes in membrane systems. The experimental work on light-driven conformational changes in bacteriorhodopsin is reviewed. To make these measurements, we developed a technique called phase modulation of fluorescence energy transfer. This allows the observation of distance changes in light-driven conformational transitions. This work showed that the location of the retinal in bacteriorhodopsin changed significantly during its photocycle (Hasselbacher *et al.,* 1986; Hasselbacher and Dewey, 1986), revealing the conformational flexibility of the protein. These results are discussed with reference to the proton transport process. Our method may be extended to membrane proteins that are not light-driven if their conformational transitions can be coupled to light-driven processes. The extension of this technique will also be discussed.

For the theoretical analysis of kinetically resolved energy transfer, two cases are considered, both of which involve energy transfer between lipids and protein. The first situation occurs when a point on the protein changes its distance relative to the membrane surface. In this case, an existing acceptor or donor merely changes locations. In the second case, a new absorbing species is created where there was not one before. This frequently occurs in light-driven biochemical processes and it may be possible to selectively transfer energy to these states. The analysis for this second case and its application to bacteriorhodopsin is also presented.

For the first situation, it is assumed that the system may be externally perturbed to drive a conformational change. By resolving the fluorescence energy transfer on the time scale of the conformational transition, it is possible, in principle, to determine changes in probe location. When the donor is located on the protein, the time course of the steady-state fluorescence in the presence and absence of donor is determined. Taking the ratio, $Q_{DA}(t)/Q_D(t)$, corrects for fluorescence changes associated with the transition which are not due to energy transfer. However, an additional correction may have to be made for changes in R_o resulting from changes in quantum yield (donor on the protein) or from changes in absorbance which affect J, the overlap integral (acceptor on the protein). From a kinetic analysis of the conformational process, it should be possible to calculate the amplitude of the change in the quantum yield ratio, $\Delta\{Q_{DA}/Q_D\}$, for the conformational change. In the case of transfer to (or from) a membrane surface this difference (designated δ) is given by:

$$\delta = \sum_i \frac{1}{1 + \sum_j [R_o'/(R_{ij}')]^6} - \frac{1}{1 + \sum_j (R_o/R_{ij})^6} \tag{13}$$

where the primed quantities represent parameters in one conformational state and the quantities without primes are for a second conformational state. It is assumed that the first state (primed) has a distance of closest approach R_e' which is shorter by a distance h than in the original state. For completeness this state is also assumed to have a different R_o. Using the geometric variables defined in the membrane model of Fig. 1, one has:

$$R_{ij}'^2 = R_{ij}^2 + h^2 - 2R_{ij}h\cos\theta_{ij} \tag{14}$$

Two series expansions are now performed. First, Eq. (13) is expanded in powers of $(R_o'/R_{ij}')^6$ and $(R_o/R_{ij})^6$. The second expansion is:

$$(R_o'/R_{ij}')^6 = (R_o'/R_{ij})^6[1 + (h^2 - 2hR_{ij}\cos\theta_{ij})/R_{ij}^2]^{-3}$$
$$= (R_o'/R_{ij})^6[1 - 3(h^2 - 2hR_{ij}\cos\theta_{ij})/R_{ij}^2 + \cdots] \tag{15}$$

This results in two series, one independent and one dependent on h. The next step is to convert from summations to integrals, giving:

$$\delta = 2\pi\sigma(R_o'^6 - R_o^6) \int_{R_e}^{\infty} R^{-6}R\,dR + \cdots$$
$$- 6\pi\sigma R_o'^6 \int_0^{\pi/2} \int_{R_e}^{\infty} \frac{(h^2 - 2hR\cos\theta)}{R^8} R\,dR\,d\theta + \cdots \tag{16}$$

In these integrations there was an implicit integration over the polar coordinate angle and this gives a factor of 2π. This is distinct from the angle θ as given in Fig. 1. The first series (h independent) represents a contribution to the energy transfer which results from changes in R_o. It is analogous to the series that is obtained in the normal two-dimensional energy transfer. This series can be expressed in the form found in Eqs. (3), (5), and (7) by replacing R_o^6 with $(R_o'^6 - R_o^6)$. In the case where there is no change in R_o, the entire series conveniently disappears. The second series (h dependent) is much better behaved and converges rapidly. In the case where R_o does not change, the leading term is obtained by performing the first integration and dropping a higher-order term:

$$\delta = 6\pi h R_o^6 \sigma / 5 R_e^5 \qquad (17)$$

Thus, to measure the displacement h from kinetic information, it is first necessary to measure the distance, R_e, using conventional applications. Equation (17) shows the quenching to be linearly related to the conformational displacement, h.

In the second case, a new absorbing species is formed transiently as a result of a perturbation on the system. This is a special, simple condition of the first case where one of the R_o's is equal to zero. In most photosynthetic systems the fraction of the population in the transient state is low (less than 10%). If the transient species is used as an acceptor, this situation represents energy transfer from multiple donors to a single acceptor (see Hasselbacher *et al.*, 1986) and Eq. (18) holds:

$$Q_{DA}/Q_D = \sigma(t)\pi R_o^6 / 4 R_e^4 \qquad (18)$$

In this case the time dependence of the acceptor's surface density is a function of the kinetic response of the system and the nature of the perturbation. In favorable situations the surface density can be varied by changing the strength of the perturbation. For these experiments, the distance of closest approach is obtained rather than the change in distance. The experimental data used to determine the changes in distances consist of quenching amplitudes associated with kinetic processes. In simple situations such as a two-state transition, the distances measured by this technique will be associated with the specific molecular species involved in the energy transfer process. However, for more complicated kinetic mechanisms, the quenching amplitudes will represent the normal modes of the reaction. The distances measured in these instances are no longer related to a single, molecular species. Instead, the appropriate linear combination of the normal quenching amplitudes is required to determine the distances to individual species. To determine such distances, a reaction mechanism must be assumed and a normal mode analysis performed so that a linear combination of amplitudes may be related to specific intermediates. If the reaction mechanism has not been established, then the interpretation of the time-resolved fluorescence energy transfer becomes model-dependent. However, the energy transfer data may be an aid in discriminating between kinetic models. First, it may show kinetic changes that could not be observed with other spectroscopic probes. More importantly, it provides physical restrictions on the nature of the amplitude changes. If a normal mode analysis of a mechanism provides a combination of amplitudes which gives unrealistically short or long distances, then the mechanism may be safely ruled out. Thus, in addition to providing unique structural information, the technique also allows a more rigorous testing of kinetic models.

The most extensive application of kinetic-resolved fluorescence energy transfer has been to bacteriorhodopsin, the light-driven proton pump from *Halobacterium halobium* (Hasselbacher *et al.,* 1986; Hasselbacher and Dewey, 1986). Bacteriorhodopsin has been extensively studied by physical techniques. It provides an unusually sturdy protein for investigating primary events in photosynthesis and ion transport. It is especially convenient for energy transfer studies. Conformational transitions may be induced by exciting the retinal chromophore into its photocycle. Since the retinal undergoes a cyclic transformation following the absorption of light, the system replenishes itself, unlike the visual pigment, rhodopsin. One of the photocycle intermediates, M_{412}, is spectrally well removed from the others so it can be used for selective energy transfer. This is a particularly important intermediate as it is closely associated with proton transport. Finally, bacteriorhodopsin can be reconstituted in monomeric form into phospholipid vesicles. This allows the easy interpretation of energy transfer from lipid donors.

The photocycle of bacteriorhodopsin has been the subject of intensive investigation and it remains an unsolved problem. It has a surprising complexity, consisting of branched and parallel pathways (see Ottolenghi, 1980). An additional controversy has been the observation of changes in stoichiometry of proton release with some techniques (Ort and Parsons, 1979) but not with others (see Grzesiek and Dencher, 1986). Recently an attempt has been made to reconcile these differences using a kinetic analysis (Sinton and Dewey, 1988). A mechanism has been proposed based on the correlation of the M state kinetics with the H^+ uptake kinetics in which the quantum yield of proton transport may vary but only one proton may be pumped. This mechanism, shown below, consists of parallel productive and nonproductive pathways:

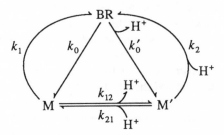

It is consistent with much of the kinetic data. Because productive and nonproductive pathways are in competition, the quantum yield for the process may vary without evoking changes in the structure of the mechanism. It also has a certain intuitive appeal. For any light-driven ion or electron transport process, initially a charge-separated state must be formed. This state has two alternatives, recombination of the charges (nonproductive path) or further separation of the charges (productive path). It is anticipated that a conformational change is required to achieve the further separation of charges. This would correspond to the interconversion step between M and M'. Estimates of the rate constants k_{12} and k_{21} put them in the range from 10 to 10^3 sec^{-1} (Sinton and Dewey, 1988). This is the same time scale required for a large scale conformational process. If the transition between the two M states is associated with a large conformational change, there may be changes in position of the retinal during this process. This was investigated using phase modulation of fluorescence energy transfer.

Phase-lifetime spectroscopy has provided a simple and powerful technique for kinetically resolving changes in the retinal position of bacteriorhodopsin during its photocycle. In this technique the photocycle is driven using periodic actinic illumination. The surface density of the M photocycle intermediates can be varied by changing the frequency of modulation of the actinic light. Simultaneous measurement of the retinal absorbance of the intermediate and the resulting fluorescence quenching due to energy transfer were made using phase-sensitive detection. This highly sensitive detection allows the measurement of surface densities as low as 10^{-8} Å^{-2} and energy transfer efficiencies as low as 10^{-4}. The modulation relaxation kinetic spectrometer used to measure photocycle intermediate and fluorescence quenching amplitudes is shown in Fig. 2. It consists of two light sources: an actinic light source, S1, that drives the bacteriorhodopsin photocycle and the probe source, S2. By adjusting the wavelength of the monochromometer, M, the probe source can be used to excite the fluorescence donor or to monitor the absorbance of an intermediate. The photomultiplier, PM1, is at right angles to both the actinic and probe beams and is used to measure the fluorescence quenching. The photomultiplier, PM2, that is colinear with the probe beam measures the absorbance of the intermediate. This design allows the measurement of both absorbance and fluorescence signals under identical conditions. A lock-in amplifier was used to measure the amplitude of the signal response. The actinic beam was chopped with a variable speed mechanical chopper which generates a reference signal at the chopping frequency. This supplied the reference frequency for the amplifier to "lock-in" to. Figure 3 shows a plot of the function $[E/(1 - E)]$ versus surface density of the M intermediate of the bacteriorhodopsin photocycle. The two curves represent probes at different membrane locations. The surface densities of the photocycle intermediate are so low that there is on the average one molecule in the M state per vesicle. Thus, energy transfer occurs from multiple donors to a

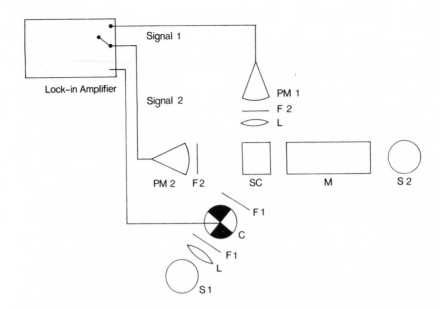

FIGURE 2. Schematic diagram of the phase-lifetime spectrophotometer used for kinetic resolution of changes in fluorescence energy transfer. Symbols are described in the text. Reprinted with permission from *Biochemistry* **25**:668–676. Copyright 1986, American Chemical Society.

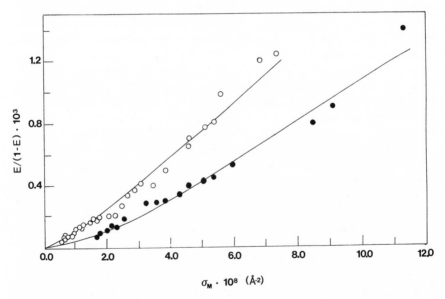

FIGURE 3. Plot of efficiency of energy transfer from surface-labeling fluorescent lipid donor to the M state retinal of bacteriorhodopsin versus surface density of bacteriorhodopsin. Reprinted with permission from *Biochemistry* **25**:6236–6243. Copyright 1986, American Chemical Society.

single acceptor. At these low surface densities and extremely low efficiency of energy transfer, Eq. (18) may be rearranged to give:

$$E/(1 - E) = \pi\sigma R_o^6/2R_e^4 \tag{19}$$

As can be seen from Fig. 3, the curves are not linear with surface density. This is a result of bacteriorhodopsin having two M intermediates that contribute to the absorbance signal. These intermediates are at different locations in the protein and therefore make different contributions to the slope of the curve. These two contributions can be deconvoluted by fitting the absorbance amplitude dispersion curve to two kinetic processes. This allows a separate determination of the surface density due to both the slow-decaying (55 msec) and fast-decaying (2 msec) intermediates. The curves in Fig. 3 are then fit by a bilinear regression using these two different surface densities as independent variables.

Using fluorescence probes in which the fluorophore is located at different depths in the phospholipid bilayer, allows a variety of distances to be measured. A schematic representation of the observed distances is given in Fig. 4. As can be seen, there is considerable movement of the retinal during three different states of the protein. A detailed theoretical analysis has been used to establish the error limits of our measurements (Hasselbacher and Dewey, 1986). The main source of error results from assumptions made about the orientational averaging of the transition dipole moments participating in the energy transfer process. Our estimated error is less than 10%. Thus, the observed distance changes are significant. These results reveal the structural dynamic nature of the bacteriorhodopsin photocycle. Because the M intermediates decay on a relatively slow time scale, the protein has time to react to the change in the electronic charge distribution of the retinal. Conformational adjustments can be made on this time scale so as to minimize the energy of the retinal–protein interaction. Thus, the protein may be "re-

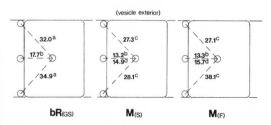

FIGURE 4. Schematic representation of distances from retinal to fluorescent lipid probes labeling bilayer surface or interior for bacteriorhodopsin in the ground state (bR_{gs}), the slow-cycling M state (M_S), and the fast-cycling M state (M_F). Distances are in angstroms and are not drawn to scale. Fluorescent donors used were: (a) Dil-C$_{18}$; (b) 16-(9-anthroyloxy)palmitic acid; (c) 2-(octadecylamino)naphthalene; (d) 1-pyrenehexadecanoic acid. Reprinted with permission from *Biochemistry* **25**:6236–6243. Copyright 1986, American Chemical Society.

laxing" into its most stable conformation in order to accommodate the retinal. This process results in effective changes in the distance of the retinal to the protein periphery (and thus the distance of closest approach of the lipid probe). These results verify that large structural differences occur between the two M intermediates as would be expected from the proposed photocycle mechanism. The specific nature of these structural differences is unclear. More detailed information may be obtained by labeling the protein at specific locations and observing energy transfer to the retinal. Such work is in progress.

This application of phase-lifetime spectroscopy provides a new structural technique with millisecond time resolution. Currently it is restricted to light-driven conformational changes. It may be extended to other membrane proteins by coupling a light-driven process to a second transport process. If the protein of interest responds to a membrane potential or pH gradient, it may be reconstituted along with bacteriorhodopsin. Conformational transitions in this protein may then be driven by light with bacteriorhodopsin's proton pump acting as the coupling mechanism. The kinetic analysis of this situation has been presented (Dewey, 1987) and early experiments have been performed on the calcium ATPase co-reconstituted with bacteriorhodopsin in phospholipid vesicles (Wu and Dewey, 1987). These experiments demonstrated that the ATP hydrolysis/synthesis equilibrium catalyzed by the ATPase could be perturbed by light-induced membrane potential changes resulting from bacteriorhodopsin's proton transport. Calcium transients were measured in this system and the relaxation kinetics of the calcium ATPase were resolved. Given a more complete understanding of the calcium transport kinetics of the ATPase, it should be possible to resolve structural transitions in this system using energy transfer. These early results suggest that it will be feasible to extend these kinetically resolved energy transfer measurements to a wide variety of membrane proteins.

4. EXCIMER FLUORESCENCE ENERGY TRANSFER

Often it is useful to observe the existence of ternary complexes or ternary phase behavior. For instance, receptor-mediated processes may require the formation of a complex between a receptor, G protein, and an enzyme. We have been developing excimer fluorescence energy transfer as a technique to approach such problems. In these cases, excimers result from the formation of a binary complex. Energy transfer from the excimer state to an appropriate acceptor will then indicate ternary complex formation.

Excimers are excited state dimers and may form when monomers are very close to each other or can rapidly diffuse together. Excimer formation has been used as a monitor of lateral distribution in lipid bilayers (Hresko *et al.*, 1986; Somerharju *et al.*, 1985; Wiener *et al.*, 1985). It has also been used to indicate the proximity of lipoic acids in multienzyme complexes (Angelides and Hammes, 1979). By observing energy transfer from the excimer state, it is possible to detect the proximity of a third moiety labeled with an energy transfer acceptor. The theoretical background for this technique may be developed using an extension of the mechanism given by Birks *et al.* (1963) for excimer formation. The decay of the excimer fluorescence in the presence of an energy acceptor is dictated by the rate processes shown in the following diagram:

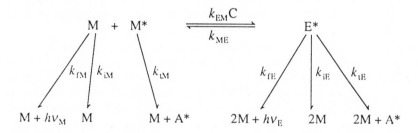

where M^*, E^*, and A^* are the excited monomer, excimer, and acceptor, respectively. The various rate constants are defined as indicated and C is the monomer concentration. The efficiency of energy transfer from the excimer, E, is defined as:

$$E = k_{tE}\{k_{tE} + k_{fE} + k_{ME} + k_{iE}\}^{-1} \qquad (20)$$

To experimentally determine this parameter, the fluorescence intensities of the excimer and monomer in the absence (F_E and F_M) and presence of acceptor (F_{EA} and F_{MA}) must be measured. The intensity ratios are:

$$F_E/F_M = (k_{fE}/k_{fM})k_{EM}C(k_{iE} + k_{fE} + k_{ME})^{-1} \qquad (21)$$

$$F_{EA}/F_{MA} = (k_{fE}/k_{fM})k_{EM}C(k_{tE} + k_{iE} + k_{fE} + k_{ME})^{-1} \qquad (22)$$

If it is assumed that all the rate constants except k_{tE} and k_{tM} are independent of acceptor concentration, then the efficiency of energy transfer from the excimer is given by:

$$E = 1 - (F_{EA}/F_{MA})/(F_E/F_M) \qquad (23)$$

For membrane systems E can now be analyzed as described in Section 2 for Eq. (6).

For cases in which R_o is to be determined independently, the quantum yield, Q_E, of the excimer donor must be known. This parameter is not as easily determined as the monomer quantum yield, Q_M. These two parameters are:

$$Q_M = k_{fM}(k_{fM} + k_{iM} + k_{EM}C)^{-1} \qquad (24)$$

$$Q_E = k_{fE}(k_{fE} + k_{iE} + k_{ME})^{-1} \qquad (25)$$

Using these quantities, it can be shown that:

$$Q_E = Q_M(F_E/F_M)[(k_{ME} + k_{EM}C)/k_{EM}C] \qquad (26)$$

The quantum yield for the monomer and the fluorescence ratio may be experimentally determined. The required rate constants are somewhat more difficult to determine and require either a closed form (Birks *et al.*, 1963) or iterative (Hresko *et al.*, 1986) analysis of steady-state and phase modulation data.

The validity of this theoretical approach was experimentally tested in a lipid vesicle system (Rundell, 1988). This was done by first measuring the distance of closest approach of pyrene monomer probes to an acceptor at the membrane surface. A variety of fatty acid probes labeled with pyrene located at different depths from the membrane surface were used. At higher pyrene concentrations, excimers are formed and energy transfer from the excimer state to the surface probe was measured. Assuming that the pyrene excimers were in the same location as the monomers and using the known surface density of acceptor, it was possible to calculate the characteristic Foerster distance for the excimer–acceptor pair. Table II shows the distance of closest approach, R_o for the monomer, and R_o for the excimer. Our results show that both the monomer and excimer R_o are independent of the specific pyrene probe used. Since the spectral properties of the different probes change very little, this result indicates that the assumptions of the model were correct.

Once the Foerster distance and membrane location of the excimer are established, it is possible to use excimer energy transfer to determine the surface density of acceptors in comparable membrane systems. This technique can be used to probe phase partitioning of the acceptor lipid. Previous work (Hresko *et al.*, 1986) established the partitioning behavior of a pyrene-

TABLE II. Excimer Energy Transfer Distances in Angstroms

Donor	R_o (monomer)	R_o (excimer)	$R_e{}^a$
1-Pyrenenonanoate	28	25	13
1-Pyrenedecanoate	30	22	13
1-Pyrenedodecanoate	28	29	18
1-Pyrenehexadecanoate	30	29	25

[a] Distance of closest approach measured to an acceptor located at the membrane surface.

labeled phospholipid (PyrPC*) in DPPC and DMPC vesicles. In the work of Rundell (1988), the surface density of DiI-C_n acceptors relative to the pyrene excimers was monitored. Comparison of the calculated surface density with the known densities could then be used to assess the phase partition behavior of these probes in mixed phase vesicles.

A simple model may be used to semiquantitatively describe phase partitioning effects on the acceptor's surface density (σ_{exp}) as determined by fluorescence energy transfer. This parameter to a first approximation (Shaklai *et al.*, 1977; Dewey and Hammes, 1980) is given by:

$$Q_{DA}/Q_D = 1 - \sigma_{exp} f(R_e, R_o) \tag{27}$$

where $f(R_e, R_o)$ is a function of the Foerster distance and the distance of closest approach. It is assumed that these parameters do not change dramatically in going from the gel to the liquid phase, so that:

$$Q_{DA}/Q_D = 1 - \sigma_{liq} f_{D,liq} f(R_e, R_o) - \sigma_{gel} f_{D,gel} f(R_e, R_o) \tag{28}$$

where σ_{liq} and σ_{gel} are the surface densities of acceptors in the liquid and gel phases, respectively, and $f_{D,liq}$ and $f_{D,gel}$ are the fraction of donors in the liquid and gel phases, respectively. Thus,

$$\sigma_{exp} = f_{D,liq} \sigma_{liq} + f_{D,gel} \sigma_{gel} \tag{29}$$

and dividing through by the overall surface density, σ, which is calculated directly from the chemical composition of the vesicles, gives:

$$\sigma_{exp}/\sigma = f_{A,liq} f_{D,liq}/f + f_{A,gel} f_{D,gel}/(1-f) \tag{30}$$

where f is the fraction of the bulk lipid in the liquid phase. Equation (30) is more conveniently used if it is expressed entirely in terms of f and the partition coefficients for donor and acceptors. The partition coefficient for the donor, P_D, is given by:

$$P_D = [D]_{liq}/[D]_{gel} = (n_{D,liq}/N_{liq})/(n_{D,gel}/N_{gel}) \tag{31}$$

where n_D represents the number of donors in the indicated phase and N is the total number of lipids in the indicated phase. An analogous definition follows for acceptors. Substituting Eq. (31) into Eq. (30), one obtains:

$$\sigma_{exp}/\sigma = (1 + P_A P_D f - f)/\{(1 + P_D f - f)(1 + P_A f - f)\} \tag{32}$$

This equation allows the observed surface density from energy transfer measurements to be related to the bulk phase behavior. This provides an approx-

---

* Abbreviations used in this chapter: PyrPC, 1-palmitoyl-2-[10-(1-pyrenyl)decanoyl]phosphatidylcholine; DMPC, dimyristoylphosphatidylcholine; DPPC, dipalmitoylphosphatidylcholine; DiI-C_n, 3,3,3',3'-tetramethylindocarbocyanine perchlorate with fatty acid chains of length n located at the 1 and 1' positions.

imate description of the effects of phase equilibrium on energy transfer. Note that the expression is symmetric with respect to donor and acceptor. Thus, without prior knowledge of the phase preference of either donors or acceptors, it would be impossible to attribute partitioning behavior specifically to one or the other. If either donor or acceptor has a partition coefficient equal to one, then σ_{exp}/σ also will equal one. Figures 5 and 6 shows the effects of different phase partitioning behavior on the surface density ratio as determined using Eq. (32) and assuming different partition coefficients. As expected, when the donors and the acceptors have the same phase preference, energy transfer is enhanced and $\sigma_{exp}/\sigma > 1$. When they have opposite preferences, this ratio can be less than one. When a single homogeneous phase exists, i.e., $f = 0$ or 1, then the ratio is one and the energy transfer experiments provide accurate determinations of the surface density of acceptors.

These theoretical curves may be compared with experimental ones in which donors and acceptors of known phase preference were used (Rundell, 1988). For investigating phase behavior one does not necessarily have to use excimer fluorescence energy transfer. However, it can be helpful because excimer formation is sensitive to the local probe concentration and can be used to monitor phase partitioning. For instance, an excimer fluorescence study (Hresko *et al.*, 1986) showed that a pyrene-labeled phosphatidyl-choline prefers the fluid phase in DPPC vesicles but has no phase preference in DMPC vesicles. This probe was then used as a donor and indocarbocyanine

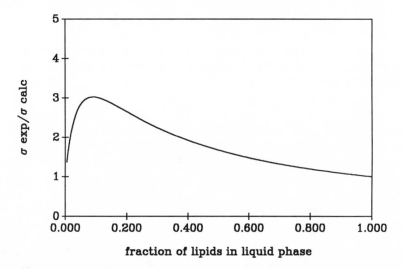

FIGURE 5. Calculated plot of σ_{exp}/σ versus f, the fraction of bulk lipid in the liquid state for $P_D = 10$ and $P_A = 10$. See text for definitions of parameters.

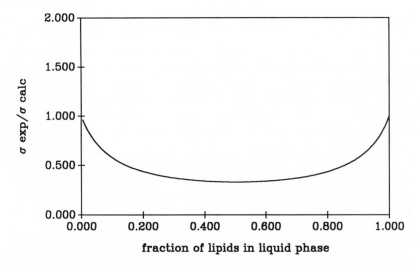

FIGURE 6. Calculated plot of σ_{exp}/σ versus f, the fraction of bulk lipid in the liquid state for $P_D = 10$ and $P_A = 0.1$. See text for definitions of parameters.

dyes of varying chain lengths were used as acceptors (Rundell, 1988). The phase preference of the acceptor was chain length dependent (Klausner and Wolf, 1980) and, therefore, provided a system of defined phase behavior. Figure 7 shows the ratio, σ_{exp}/σ, versus temperature for pyrene excimers transferring energy to DiI-C_{12}, DiI-C_{18}, and DiI-C_{20} in DPPC. The C_{18} and

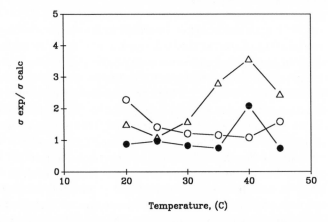

FIGURE 7. Experimental plots of σ_{exp}/σ versus temperature in DPPC vesicles. The donor was pyrenedecanoate excimers and the acceptors were DiI-C_{12} (O), DiI-C_{18} (●), and DiI-C_{20} (△).

C_{20} probes prefer the fluid phase while the C_{12} probe has no preference for either phase (Klausner and Wolf, 1980). As the theoretical model would predict, the surface density ratio increases when donor–acceptor pairs have like preferences ($P_D > 1$ and $P_A > 1$). In temperature regions where both fluid and gel phases (25–40°C) coexist, there is an enhancement of energy transfer ($\sigma_{exp}/\sigma > 1$) as a result of both donors and acceptors being concentrated in the liquid regions. When the acceptor shows no phase preference, the surface density ratio remains constant at one. Both results are consistent with predictions of the simple theory.

In this section, the theoretical formalism for analyzing fluorescence energy transfer from excimer states was presented. This formalism was tested by experimental work in model vesicle systems. This technique has a variety of potential applications and can be used to monitor the formation of ternary complexes. This should provide a means of investigating the oligomeric structure of membrane proteins and of protein–protein interactions in signaling processes. The key to the further application of this technique is the ability to label moieties so the label is accessible for excimer formation. This requires either a long, flexible attachment or a labeling site near the protein–protein contact site. Once excimer-forming binary complexes can be created, it is straightforward to observe the degree of energy transfer to a third location.

5. EXCITON FLUORESCENCE ENERGY TRANSFER: TRYPTOPHAN IMAGING REVISITED

Frequently the situation arises in which energy transfer may be observed from more than one intrinsic donor on a protein. In such cases it would be helpful to put physical constraints on the locations of the donors. To achieve this goal, Kleinfeld developed a technique which he referred to as "tryptophan imaging" (Kleinfeld, 1985). In this method a Monte Carlo calculation is performed to determine the tryptophan distributions which are consistent with the efficiency of energy transfer. This analysis was used to investigate the structure of cytochrome b_5 (Kleinfeld and Lukacovic, 1985) and is potentially applicable to a wide range of systems. An implicit assumption in this model is that fluorescence energy transfer from each donor is an independent process. This assumption breaks down in two situations. First, we consider the possibility of self-transfer between donors. Foerster distances for self-transfer are usually small and most applications have extremely dilute concentrations of donors. However, in cases such as multiple tryptophans in a single protein, it is possible that donors are very close to one another and even a small R_0 for self-transfer can have a large effect on the energy transfer processes. A second, more extreme situation is when donors are close enough to have strong dipolar coupling. This results in the formation of a delocalized exciton state. If this occurs, the donor wavefunctions are no longer independent and the value of the distances measured by energy transfer will be intermediate to the individual donors' distances.

First the case of donors which can self-transfer is treated. For simplicity, two donors embedded in a protein are considered. The reaction mechanism for fluorescent decay is given by:

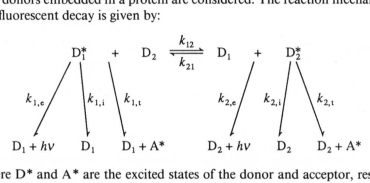

where D^* and A^* are the excited states of the donor and acceptor, respectively, and the subscript number refers to a specific donor location. The rate constants are defined as in the diagram with k_e representing a radiative decay constant and k_i is the nonradiative process. The remaining constants are all energy transfer constants which take the form of Eq. (2). The constants k_{12} and k_{21} need not be equal as the R_0 for self-transfer can change due to difference in quantum yields and R_0 for a given donor. The efficiency of energy transfer for two donors may be defined as follows (Kleinfeld, 1985):

$$E = 1 - (I_1 + I_2)/(I_1^0 + I_2^0) \tag{33}$$

where I and I^0 represent the fluorescence intensity in the presence and absence of donors. This is related to a sum of efficiencies representing contributions from each donor:

$$E = \alpha_1 E_1 + \alpha_2 E_2 \tag{34}$$

where

$$\alpha_1 = \epsilon_1 Q_1/(\epsilon_1 Q_1 + \epsilon_2 Q_2); \qquad \alpha_2 = \epsilon_2 Q_2/(\epsilon_1 Q_1 + \epsilon_2 Q_2)$$

with ϵ being the extinction coefficient and Q the quantum yield. In previous work the efficiency distribution was determined by Monte Carlo calculations and an average quantum yield was assumed for each donor (Kleinfeld and Lukacovic, 1985). The effect of introducing self-transfer is that the quantum yields are not independent quantities. In the previous formalism, highly quenched donors, i.e., donors with large k_i, did not contribute to energy transfer and could be ignored. They were treated by reducing the number of donors by one when calculating the average quantum yield. If self-transfer is added, highly quenched molecules have a more profound effect. They can act as a trap for energy, reducing the quantum yields of neighboring donors that transfer energy to them. Since they provide little fluorescence energy transfer to acceptors, the efficiency of the entire system may be greatly reduced. This effect may be seen more quantitatively by examining the quantum yields which enter into the weighted average of Eq. (34):

$$Q_1 = k_{1,e}/(k_{1,e} + k_{1,i} + k_{12}) = Q_1'(1 - E_1') \tag{35}$$

where $Q_1' = k_{1,e}/(k_{1,e} + k_{1,i})$ which is the quantum yield if self-transfer did not occur and $E_1' = k_{12}/(k_{1,e} + k_{1,i} + k_{12})$ which is the efficiency of energy transfer between donors. An analogous equation to Eq. (35) holds for the quantum yield of the second donor. Substituting Eq. (35) into (34) gives the following expression:

$$\alpha_1 = \epsilon_1 Q_1' \{ \epsilon_1 Q_1' + \epsilon_2 (R_{12}^6 + R_{1,o}'^6)/(R_{12}^6 + R_{2,o}'^6) \}^{-1} \tag{36}$$

where R_{12} is the distance between donors and $R_{i,o}'^6$ is the Foerster distance for self-transfer from the ith donor. Using Eqs. (34) and (36) it is instructive to compare the effect of self-transfer on the weighting of the individual efficiencies. Assuming a situation where $Q_1' = 0.9$ and $Q_2' = 0.1$ and $\epsilon_1 = \epsilon_2$, if no self-transfer occurs, $\alpha_1 = 0.9$ and $\alpha_2 = 0.1$. Now, let $R_{12} = R_{1,o}'$. If the only difference between the two Foerster distances, $R_{1,o}'$ and $R_{2,o}'$, is due to the quantum yield, then $R_{2,o}' = 0.693\, R_{1,o}'$. This results in the $\alpha_1 = 0.83$ and $\alpha_2 = 0.17$, giving a large increase in the contribution from the weaker of the two donors. Thus, self-transfer via a Foerster mechanism may have a profound effect upon the observed efficiency.

Using the equations developed here, it is possible to introduce self-transfer effects into the Monte Carlo calculations used in "tryptophan imaging." In the original application of this technique (Kleinfeld and Lukacovic, 1985), the distribution of transfer efficiencies was used to put crude constraints upon the values for the quantum yields. This procedure did not introduce too great an uncertainty because the model was more sensitive to the individual efficiencies than the individual quantum yields. With the self-transfer model, this is not always the case and a more detailed knowledge of the quantum yields is a requirement. Thus, it would appear to be extremely difficult to experimentally unravel all the required parameters. Nevertheless, it may still be possible to perform a double Monte Carlo calculation in which the distribution of efficiencies and of quantum yields is probed.

The second situation under consideration is the case in which the donors are brought close enough to form exciton states. When this occurs, the donor wavefunction has actually changed. The exciton wavefunction, Ψ_g, for the ground state of a donor dimer is given by $\Psi_g = \phi_1\phi_2$ (see Tinoco, 1970), where ϕ_1 and ϕ_2 are the ground state molecular wavefunctions of the donors at positions 1 and 2, respectively. Two, nondegenerate exciton excited state wavefunctions exist. They are given by:

$$\Psi_+ = (\phi_1'\phi_2 + \phi_1\phi_2')/\sqrt{2} \tag{37}$$

$$\Psi_- = (\phi_1'\phi_2 - \phi_1\phi_2')/\sqrt{2}$$

where the prime designates the excited states of the molecular wavefunctions. The energies of these states depend on the distance and orientation between the dipole moments of the donors. These new wavefunctions are now used in the Foerster theory to calculate the rate constants for energy transfer, k_T. It is now possible to transfer fluorescence energy out of either of the two exciton states. The goal of this exercise is to establish the distance dependence of these rate expressions. The rate constant for each of the exciton states is given by:

$$k_{T\pm} \propto |\langle \Psi_\pm \phi_A | V | \Psi_g \phi_A' \rangle|^2 \tag{38}$$

where $\pm$ refers to the rate of transfer from the respective exciton states and V is the dipolar interaction potential between donor(s) and acceptor. The acceptor wavefunction is designated by the subscript A. The interaction potential takes the form:

$$V = \kappa |\mu_{D1}| |\mu_A|/R_1^3 + \kappa |\mu_{D2}| |\mu_A|/R_2^3 \tag{39}$$

where the μ's are the transition dipole moments and $R_{1(2)}$ is the distance from donor 1 (2) to the acceptor. For simplicity we now assume symmetric donors. Inserting Eq. (39) into (38) gives:

$$k_{T\pm} = (R_o^6/\tau_D)(R_1^{-6} + R_2^{-6} \pm R_1^{-3}R_2^{-3}) \tag{40}$$

where R_o is the Foerster distance obtained for the monomer case. Thus, the energy transfer from each excimer state occurs at a different rate due to its different functional dependence on distance. The observed overall energy transfer takes the same form as Eq. (34). Given the symmetric model ($Q_1 = Q_2$), one might expect that the individual exciton efficiencies would add to give a form identical to two independent monomeric donors. Even in this simplifying case, the overall efficiency does not reduce to a simple expression. This is because the weighting factors, α_i, contain the extinction coefficients, ϵ_i. Excitonic interactions result in a splitting of the monomeric absorbance spectrum so that the exciton states will absorb maximally at a different wavelength. In general, electronic transitions to these states will have different dipolar strengths and consequently, the extinction coefficient will be different. As a result of the broad absorbance spectra, it is difficult to experimentally determine parameters characterizing the excitonic interactions. Thus, this situation also is difficult to analyze with existing experimental limitations. Unfortunately, when molecules are very close to one another, excitonic interactions almost certainly occur. One experimental check to assess if such effects are occurring is to measure the fluorescence energy transfer efficiency as a function of excitation wavelength. A dependence on wavelength indicates that the different exciton states are being preferentially populated at different wavelengths.

This section serves to demonstrate the pitfalls that may occur when considering energy transfer from multiple donors in close proximity. For tryptophan energy transfer the R_o for self-transfer should be very small. However, if the tryptophans are within a couple of residues of each other there will doubtless be some self-transfer. If tryptophans are on adjacent residues, they will likely demonstrate excitonic interactions. These situations prohibit a realistic analysis of experimental data. Site-directed mutagenesis could be of great assistance in better defining these systems and their problems. However, changes of residues near a fluorophore can always affect the quantum yield for reasons other than self-transfer or excitonic interactions, so that even structurally safe mutations may give misleading changes in spectral properties.

6. PERSPECTIVES

In recent years, fluorescence energy transfer has been applied to increasingly complex topographical and temporal situations. Parallel theoretical and experimental developments have contributed to these applications. Much of this development has been prompted by the difficulty in applying conventional structural techniques to membrane systems. These efforts have made it possible to quantitatively probe the details of membrane processes. The theoretical advances of Fayer and co-workers suggest that quantitative descriptions of complicated energy transport processes may be possible. This work demonstrates the utility of time-resolved fluorescence depolarization for investigating multiple donor–acceptor distributions. It is anticipated that this technique will find increased application for energy transfer studies of biological systems.

Applications of fluorescence energy transfer to protein conformational dynamics offer exciting possibilities. Few specific structural techniques provide the kinetic resolution that allows a visualization of transient conformations of biomolecules. While the absolute error in distances measured by energy transfer is usually on the order of 10%, the error in differences in distances will be much less if the characteristic Foerster distance does not change. Nevertheless, most successful applications will be on systems which show distance changes of 5 Å or greater. The kinetic resolution of energy transfer will doubtless find its most productive applications when applied to multisubunit complexes. These proteins can have large displacements between sites on adjacent subunits and are not accessible to traditional structural techniques.

While this chapter has emphasized the quantitative aspects of fluorescence energy transfer in membrane systems, more qualitative experiments have found increased utility in cell biological applications. Energy transfer is a convenient tool for monitoring intracellular transport and fusion processes. The observation at a cellular level of quenching due to mixing of a population of donors with acceptors can easily be accomplished with fluorescence microscopy. Thus, it is possible to observe the trafficking of membrane components *in vivo*. Likewise, monitoring ternary complex formation with excimer energy transfer may be accomplished *in vivo* and need not be approached quantitatively. If the acceptor is fluorescent, sensitized emission can be a simple indicator of the existence of such complexes. Energy transfer cascades may also be used to observe higher-order complexes. Using successive acceptors as donors for the next acceptor, simple questions of the degree of association of multicomponent systems can be approached. The main problem in these applications is the protein chemistry required for specific labeling. Using such approaches, fluorescence energy transfer offers a means of tackling increasingly complex problems in membrane biochemistry and cell biology.

7. REFERENCES

Angelides, K., and Hammes, G. G., 1979, *Biochemistry* **18**:1223–1229.
Arvinte, T., Wahl, P., and Nicolau, C., 1987, *Biochemistry* **26**:765–772.
Aso, Y., Kano, K., and Matsuo, T., 1980, *Biochim. Biophys. Acta* **599**:403–416.
Baird, B. A., Pick, U., and Hammes, G. G., 1979, *J. Biol. Chem.* **254**:3818–3825.
Birks, J. B., Dyson, D. J., and Munro, I. H., 1963, *Proc. R. Soc. London Ser A* **275**:575–588.
Blumberg, W. E., 1985, *NATO ASI Ser., Ser. A* **71**:95–122.
Chan, S. S., Arndt-Jovin, D. J., and Jovin, T. M., 1979, *J. Histochem. Cytochem.* **27**:56–64.
Chatelier, R. C., Rogers, P. J., Ghiggino, K. P., and Sawyer, W. H., 1984, *Biochim. Biophys. Acta* **776**:75–82.
Chejanovsky, N., Evtan, G. D., and Loyter, A., 1984, *FEBS Lett.* **174**:304–309.
Cheng, K. H., Wiedmer, T., and Sims, P. J., 1985, *J. Immunol.* **135**:459–464.
Dangreau, H., Joniau, M., De Cuyper, M., and Hanssens, I., 1982, *Biochemistry* **21**:3594–3598.
Davenport, L., Dale, R. E., Bisby, R. H., and Cundall, R. B., 1985, *Biochemistry* **24**:4097–4108.
Dewey, T. G., 1987, *Biophys. J.* **51**:809–815.
Dewey, T. G., and Datta, M., 1989, *Biophys. J.* **56**:415–420.
Dewey, T. G., and Hammes, G. G., 1980, *Biophys. J.* **32**:1023–1036.
Dexter, D. L., 1953, *J. Chem. Phys.* **21**:836–849.

Dissing, S., Jesaitis, A. J., and Fortes, P. A. G., 1979, *Biochim. Biophys. Acta* **553**:66–83.
Ediger, M. D., and Fayer, M. D., 1983, *J. Chem. Phys.* **78**:2518–2524.
Ediger, M. D., and Fayer, M. D., 1984, *J. Phys. Chem.* **88**:6108–6116.
Eisinger, J., and Flores, J., 1982, *Biophys. J.* **37**:6–7.
Estep, T. N., and Thompson, T. E., 1979, *Biophys. J.* **26**:195–207.
Fagan, M. H., and Dewey, T. G., 1986, *J. Biol. Chem.* **261**:3654–3660.
Fleming, P. J., Koppel, D. E., Lau, A. L. Y., and Strittmatter, P., 1979, *Biochemistry* **18**:5458–5464.
Foerster, T., 1949, *Z. Naturforsch. A* **4**:321–323.
Foerster, T., 1965, *Mod. Quant. Chem.* **3**:93–117.
Fung, B. K., and Stryer, L., 1978, *Biochemistry* **17**:5241–5248.
Gibson, G. A., and Loew, L. M., 1979, *Biochem. Biophys. Res. Commun.* **88**:135–140.
Gordon, R. G., 1968, *J. Math. Phys.* **9**:655–663.
Graue, C., and Klingenberg, M., 1979, *Biochim. Biophys. Acta* **546**:539–550.
Grzesiek, S., and Dencher, N. A., 1986, *FEBS Lett.* **208**:337–342.
Grochanour, C. R., Andersen, H. C., and Fayer, M. D., 1979, *J. Chem. Phys.* **70**:4254–4271.
Gutierrez-Merino, C., Munkonge, F., Mata, A. M., East, J. M., Levinson, B. L., Napier, R. M., and Lee, A. G., 1987, *Biochim. Biophys. Acta* **897**:207–216.
Hahn, L.-H. E., and Hammes, G. G., 1978, *Biochemistry* **17**:2423–2429.
Haigh, E. A., Thulborn, K. R., and Sawyer, W. H., 1979, *Biochemistry* **18**:3525–3532.
Hammes, G. G., 1981, *Protein–Protein Interactions,* Wiley, New York, pp. 257–287.
Hasselbacher, C. A., and Dewey, T. G., 1986, *Biochemistry* **25**:6236–6243.
Hasselbacher, C. A., Street, T. L., and Dewey, T. G., 1984, *Biochemistry* **23**:6445–6452.
Hasselbacher, C. A., Preuss, D. K., and Dewey, T. G., 1986, *Biochemistry* **25**:668–676.
Herman, B. A., and Fernandez, S. M., 1982, *Biochemistry* **21**:3275–3283.
Heyn, M. P., Blume, A., Rehorek, M., and Dencher, N. A., 1981a, *Biochemistry* **20**:7109–7115.
Heyn, M. P., Cherry, R. J., and Dencher, N. A., 1981b, *Biochemistry* **20**:840–849.
Highsmith, S., and Cohen, J. A., 1987, *Biochemistry* **26**:154–161.
Holowka, D., and Baird, B., 1983, *Biochemistry* **22**:3466–3474.
Hresko, R. C., Sugar, I. P., Barenholz, Y., and Thompson, T. E., 1986, *Biochemistry* **25**:3813–3823.
Hyono, A., Kuriyama, S., and Masui, M., 1985, *Biochim. Biophys. Acta* **813**:111–116.
Isaacs, B. S., Husten, E. J., Esmon, C. T., and Johnson, A. E., 1986, *Biochemistry* **25**:4958–4969.
Kampmann, L., 1977, *Biophys. Struct. Mech.* **3**:239–257.
Kano, K., Yamaguchi, T., and Matsuo, T., 1980, *J. Phys. Chem.* **84**:72–76.
Kano, K., Kawazumi, H., and Ogawa, T., 1981, *J. Phys. Chem.* **85**:2998–3003.
Keller, P. M., Person, S., and Snipes, W., 1977, *J. Cell Sci.* **28**:167–177.
Kirchanski, S., and Branton, D., 1980, *Proc. Annu. Meet. Electron Microsc. Soc. Am. 38th,* pp. 756–759.
Klafter, J., and Blumen, A., 1984, *J. Chem. Phys.* **80**:875–877.
Klausner, R. D., and Wolf, D. E., 1980, *Biochemistry* **19**:6199–6203.
Kleinfeld, A. M., 1985, *Biochemistry* **24**:1874–1882.
Kleinfeld, A. M., and Lukacovic, M. F., 1985, *Biochemistry* **24**:1883–1890.
Kometani, T., Kinosita, K., Jr., Furuno, T., Kouyama, T., and Ikegami, A., 1987, *Biophys. J.* **52**:509–517.
Koppel, D. E., Fleming, P. J., and Strittmatter, P., 1979, *Biochemistry* **18**:5450–5457.
Kouyama, T., Kinosita, K., Jr., and Ikegami, A., 1983, *J. Mol. Biol.* **165**:91–107.
Kunitake, T., Shimomura, M., Hashiguchi, Y., and Kawanaka, T., 1985, *J. Chem. Soc. Chem. Commun.* **12**:833–835.

MacDonald, R. I., and MacDonald, R. C., 1983, *Biochim. Biophys. Acta* **735**:243–251.

Mani, J. C., Dornand, J., and Mousseron-Canet, M., 1975, *Biochimie* **57**:629–635.

Miller, R. J. D., Pierre, M., and Fayer, M. D., 1983, *J. Chem. Phys.* **78**:5138–5146.

Morris, S. J., and Bradley, D., 1984, *Biochemistry* **23**:4642–4650.

Morris, S. J., Suedhof, T. C., and Haynes, D. H., 1982, *Biochim. Biophys. Acta* **693**:425–436.

Nagata, I., Li, R., Banks, E., and Okamoto, Y., 1983, *Macromolecules* **16**:903–905.

Nakajima, M., Yoshimoto, R., Irimura, T., and Osawa, T., 1979, *J. Biochem.* **86**:583–586.

Nakashima, N., Kimizuka, N., and Kunitake, T., 1985, *Chem. Lett.* **12**:1817–1820.

Nakashima, N., Ando, R., and Kunitake, T., 1987, *Bull. Chem. Soc. Jpn.* **60**:1967–1973.

Oesterhelt, D., Schreckenbach, T., and Walckhoff, B., 1981, *Comm. Eur. Communities,* [*Rep.*] *EUR, EUR 7591.*

Omata, Y., Aibara, K., and Ueno, Y., 1987, *Biochim. Biophys. Acta* **912**:115–123.

Ort, D. R., and Parsons, W. W., 1979, *Biophys. J.* **25**:341–354.

Ottolenghi, M., 1980, *Adv. Photochem.* **12**:97–200.

Peerce, B. E., and Wright, E. M., 1986, *Proc. Natl. Acad. Sci. USA* **83**:8092–8096.

Peters, R., 1971, *Biochim. Biophys. Acta* **233**:465–468.

Rehorek, M., Dencher, N. A., and Heyn, M. P., 1983, *Biophys. J.* **43**:39–45.

Rogers, J., Lee, A. G., and Wilton, D. C., 1979, *Biochim. Biophys. Acta* **552**:23–37.

Rundell, K. A., 1988, Master's thesis, University of Denver.

Shaklai, N., Yquaribide, J., and Ranney, H. M., 1977, *Biochemistry* **16**:5585–5592.

Shimomura, M., Hashimoto, H., and Kunitake, T., 1982, *Chem. Lett.* **8**:1285–1288.

Sims, P. J., 1984, *Biochemistry* **23**:3248–3260.

Sinton, M. H., and Dewey, T. G., 1988, *Biophys. J.* **53**:153–162.

Sklar, L. A., Miljanich, G. P., Bursten, S. L., and Dratz, E. A., 1979, *J. Biol. Chem.* **254**:9583–9591.

Smith, C. M., Satoh, K., and Fork, D. C., 1986, *Plant Physiol.* **80**:843–847.

Snyder, B., and Freire, E., 1982, *Biophys. J.* **40**:137–148.

Somerharju, P. J., Virtanen, J. A., Edlund, K. K., Vainio, P., and Kinnunen, P. K. J., 1985, *Biochemistry* **24**:2773–2781.

Struck, D. K., Hoekstra, D., and Pagano, R. E., 1981, *Biochemistry* **20**:4093–4099.

Stryer, L., 1978, *Annu. Rev. Biochem.* **47**:819–846.

Stryer, L., and Haugland, R. P., 1967, *Proc. Natl. Acad. Sci. USA* **98**:719–726.

Stryer, L., Thomas, D. D., and Meares, C. F., 1982, *Annu. Rev. Biophys. Bioeng.* **11**:203–222.

Takami, A., and Mataga, N., 1987, *J. Phys. Chem.* **91**:618–622.

Talbot, J. C., Faucon, J. F., and Dufourcq, J., 1987, *Eur. Biophys. J.* **15**:147–157.

Tamai, N., Yamazaki, T., Yamazaki, I., and Mataga, N., 1986, *Springer Ser. Chem. Phys.* **46**:449–453.

Tamai, N., Yamazaki, T., Yamazaki, I., Mizuma, A., and Mataga, N., 1987, *J. Phys. Chem.* **91**:3503–3508.

Tasaki, I., Warashina, A., and Pant, H., 1976, *Biophys. Chem.* **4**:1–13.

Tinoco, I., 1970, *Methods Biochem. Anal.* **18**:81–203.

Tron, L., Szollosi, J., and Damjanovich, S., 1987, *Immunol. Lett.* **16**:1–9.

Trung Le Doan, Takasugi, M., Aragon, I., Boudet, G., Montenay-Garestier, T., and Helene, C., 1983, *Biochim. Biophys. Acta* **735**:259–270.

Tweet, A. G., Bellamy, W. D., and Gaines, G. L., 1964, *J. Chem. Phys.* **41**:2068–2077.

Uster, P. S., and Pagano, R. E., 1986, *J. Cell Biol.* **103**:1221–1234.

Vanderkooi, J. M., Ierokamas, A., Nakamura, H., and Martonosi, A., 1977, *Biochemistry* **16**:1262–1267.

Vanderwerf, P., and Ullman, E. F., 1980, *Biochim. Biophys. Acta* **596**:302–314.

Veatch, W., and Stryer, L., 1977, *J. Mol. Biol.* **113**:89–102.

Weber, G., and Daniel, E., 1966, *Biochemistry* **5**:1900–1907.
White, T. E., and Dewey, T. G., 1987, *Membr. Biochem.* **7**:67–72.
Wiener, J. R., Pal, R., Barenholz, Y., and Thompson, T. E., 1985, *Biochemistry* **24**:7651–7658.
Wolber, P. K., and Hudson, B. S., 1979, *Biophys. J.* **28**:197–210.
Womersley, C., Uster, P. S., Rudolph, A. S., and Crowe, J. H., 1986, *Cryobiology* **23**:245–255.
Wu, X. L., and Dewey, T. G., 1987, *Biochemistry* **26**:6914–6918.

Chapter 8

Imaging Membrane Organization and Dynamics

Richard A. Cardullo, Robert M. Mungovan, and David E. Wolf

1. INTRODUCTION

In this chapter, we will discuss how the conventional techniques of nonradiative fluorescence resonance energy transfer (FRET) and fluorescence recovery after photobleaching (FRAP) can be extended, by coupling these techniques to low-light-level video microscopy and digital image processing, to provide an additional dimension of information. That is, what new information can be gained when, rather than integrating the fluorescence signal over a photomultiplier tube, one instead retains the spatial information by using a two-dimensional video detector? In this chapter, we wish to emphasize the conceptual aspects of these experiments rather than focusing on specific issues of instrumentation and software. For a discussion of these issues the reader is referred to several comprehensive texts on this subject (Castleman, 1979; Inoue, 1986; Makovski, 1983; Taylor and Wang, 1989). In our discussion, we wish also to emphasize how certain empirical manipulations of experimental data can greatly simplify complicated analysis without sacrificing either validity or accuracy.

RICHARD A. CARDULLO, ROBERT M. MUNGOVAN, and DAVID E. WOLF
• Worcester Foundation for Experimental Biology, Shrewsbury, Massachusetts 01545.

2. FRET

2.1. Basic Concepts

The technique of FRET provides spatial information with Ångstrom resolution between appropriate pairs of donor and acceptor fluorophores and as such has been called the "spectroscopic ruler." We will here discuss the concept and manifestations of FRET particularly as it relates to the study of molecular distributions in membranes. Our discussion follows the conventions and theory developed by Fung and Stryer (1978). For more thorough discussion the reader is referred to accompanying chapters by Dewey and Erickson *et al.*

The concept of FRET is shown in Fig. 1. We have two molecules, the donor, D, and an acceptor, A. At the top, we see the absorbance and the emission spectra A(D), A(A), E(D), and E(A) of donor and acceptor. Of importance is the area of overlap between the donor emission and acceptor absorbance spectra. This area is the overlap integral. Notice that if we were to excite at wavelength I, light would be emitted at wavelength II by the donor but not at wavelength III by the acceptor, because the acceptor does not absorb light of wavelength I. One could imagine a radiative phenomenon where light is absorbed by donor at I and emitted by donor at II; reabsorbed by acceptor at II and emitted by acceptor at III. *Such radiative transfer does not occur at reasonable concentrations.* What does occur is a nonradiative transfer process shown in Fig. 2. A donor molecule absorbs the photon whose electric field vector is shown by E. The excited state of the donor is

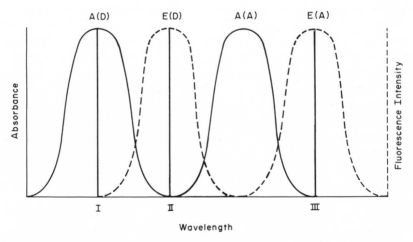

FIGURE 1. Hypothetical spectra showing overlap necessary for FRET. I = donor excitation wavelength; II = donor emission wavelength; III = acceptor emission wavelength.

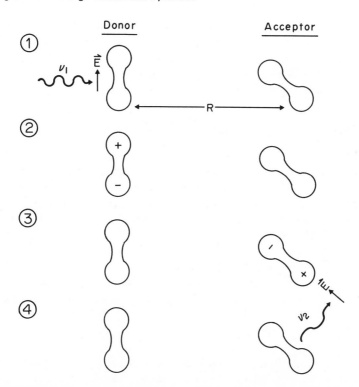

FIGURE 2. Schematic diagram of FRET showing the process of dipole-induced dipole interaction which results in nonradiative fluorescence resonance energy transfer. (1) Incident light of frequency ν_1 is absorbed and (2) induces a dipole on the donor. Rather than emitting a photon, the excited donor (3) induces a dipole on the acceptor which (4) emits a photon at frequency ν_2 at a different polarization from the incident light.

shown as a dipole with positive charge on one side and negative charge on the other. If an acceptor molecule is nearby, an oppositely charged dipole is induced on it (it is raised to an excited state). Such dipole-induced dipole interactions fall off inversely as the sixth power of donor–acceptor molecular distance. Classically, partial energy transfer can occur. Quantum mechanically, it is an all-or-nothing affair. All of the energy must be transfered, and transfer can only occur when the energy levels (i.e., the spectra) overlap. When A leaves its excited state, the emitted light is rotated or depolarized with respect to the incident light. Thus, FRET manifests itself as: (1) a decrease or quenching of donor fluorescence intensity at II, (2) an increase in acceptor fluorescence intensity at III, and (3) a depolarization of the fluorescence relative to the incident light.

A further manifestation of FRET is in the excited state lifetime. One can envision fluorescence as an equilibrium process, where how long on average a molecule stays in its excited state is a competition between the rate at which it is being driven into this state by the incident light and the sum of the rates driving it out of this state, namely fluorescence and nonradiative processes. If one adds a new nonradiative process to the system, namely FRET, leaving all else unchanged, then one favors decay which results in the shortening of the donor lifetime at II.

2.2. Basic Theory as Applied to Membranes

FRET is a technique that has been widely used to measure distances between sites on proteins, nucleic acids, and in membranes (see, e.g., Baird and Holowka, 1988; Cardullo *et al.*, 1988; Fung and Stryer, 1978; Koppel *et al.*, 1979; Kleinfeld, 1988). Solutions for generalized distributions in membranes exist (Dewey and Hammes, 1988; Fung and Stryer, 1978; Koppel *et al.*, 1979), including those which predict transfer between identical molecules in terms of fluorescence depolarization (Snyder and Friere, 1982). For purposes of illustration we will consider the treatment of Fung and Stryer (1978) for transfer where the distributions are random and where the donor and acceptor are different molecular species. It should also be pointed out that modern Monte Carlo techniques exist which enable solution of analytically intractable problems.

If one has a donor molecule whose fluorescence emission spectrum overlaps the absorbance spectrum of a fluorescent acceptor molecule, then they will exchange energy between one another by a *nonradiative* dipole-induced dipole interaction as described above. This transfer will manifest itself both by quenching of donor fluorescence in the presence of acceptor and in the sensitized emission of acceptor fluorescence. The rate of this transfer for *isolated* donor–acceptor pairs is given by

$$k_T = \frac{1}{\tau_0} \left(\frac{R_0}{R} \right)^6 \tag{1}$$

where τ_0 is the excited state lifetime of donor in the absence of acceptor, R_0 is the distance of half transfer, and R is the distance between donor and acceptor. R_0 in Ångstroms is given by

$$R_0 = (J\kappa^2 Q_0 \eta^{-4})^{1/6} \times 9.79 \times 10^3 \tag{2}$$

where J is the overlap integral in $cm^3\ M^{-1}$, κ^2 is the dipole–dipole orientation factor, Q_0 is the quantum efficiency of donor in the absence of acceptor, and η is the index of refraction of the medium.

Suppose that at time $t = 0$ a donor molecule goes into the excited state. Then the probability that it will still be in its excited state at time t in the *absence* of acceptor is

$$p_0(t) = e^{-t/\tau_0} \tag{3}$$

In the presence of acceptor this becomes

$$p(t) = e^{-(1/\tau_0 + k_T)t} \tag{4}$$

When one has a random distribution of donor–acceptor pairs, it can be shown that (4) becomes

$$p(t) = e^{-[t/\tau_0 + \sigma S(t)]} \tag{5}$$

where

$$S(t) = \int_a^{+\infty} [1 - e^{-(t/\tau_0)(R_0/r)^6}] 2\pi R \, dR \tag{6}$$

and where σ is the surface density of acceptors and a is the distance of closest approach between donors and acceptors.

The transfer efficiency is given by

$$E_T = 1 - (FI_{D,A}/FI_D) \tag{7a}$$

or,

$$E_T = 1 - \int_0^\infty p(t) \, dt \bigg/ \int_0^\infty p_0(t) \, dt \tag{7b}$$

where $FI_{D,A}$ is the steady-state fluorescence in the presence of both donor and acceptor and FI_D is the steady-state fluorescence in the absence of acceptor. For computational purposes, Eq. (6) and therefore Eqs. (7a) and (7b) are readily solved numerically.

3. FRAP

3.1. Basic Concepts

The basic principles of the technique of spot FRAP for measuring the lateral diffusibility of membrane compounds are shown in Fig. 3. Light from an ion laser is brought, using a modified fluorescence microscope, to a small spot (of typical radius of 1 μm) on the membrane, which has been tagged fluorescently. It is important that the fluorescent label be non-cross-linking and irreversibly photobleachable. The fluorescence from the spot is monitored and is found to be more or less constant. The incident light is momentarily increased 10^3- to 10^4-fold which causes a significant fraction of the fluorescence to be bleached. As a result, when the incident light is returned to the monitoring level the fluorescence is found to be greatly diminished. Typically one aims for about 70% bleaching. If there is no freedom of motion of molecules in and out of the spot, the fluorescence intensity will remain at this

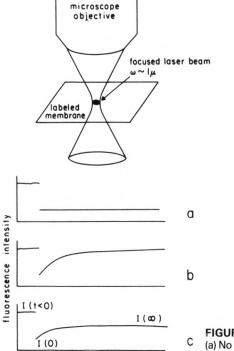

FIGURE 3. Schematic diagram of spot FRAP. (a) No recovery; (b) complete recovery; (c) partial recovery.

reduced level *ad infinitum* as shown in Fig. 3a. If, as is illustrated in Fig. 3b, there is complete freedom of motion of material in and out of the spot, say by diffusion, then the fluorescence intensity will gradually regain its initial value and the diffusion coefficient can be calculated from the half-time for recovery (Axelrod *et al.*, 1976) or by nonlinear least-squares fitting to the diffusion theory given below. An intermediate situation may exist, as illustrated in Fig. 3c, where only some of the labelled molecules are free to diffuse, in which case only partial recovery of fluorescence is equal to the fraction of all labeled molecules free to diffuse. *Thus, FRAP measures two parameters: the diffusion coefficient, D, and the fraction free to diffuse, R.* Recovery by modes other than diffusion such as unidirectional flow can also be studied by FRAP (Axelrod *et al.*, 1976).

3.2. Basic Theory

In this section we will develop and consider the theory of FRAP. Let us call the concentration profile on the membrane immediately after the bleach

$C(x, y, 0)$ and at some time later as $C(x, y, t)$. $C(x, y, t)$ is what we must determine. It is the solution to the diffusion equation:

$$\frac{\partial C(x, y, t)}{\partial t} = D\left(\frac{\partial^2 C(x, y, t)}{\partial x^2} + \frac{\partial^2 C(x, y, t)}{\partial y^2}\right) \tag{8}$$

A convenient way to solve such equations is by the method of Fourier transforms. These are discussed in greater detail in the Appendix. In general, the two-dimensional Fourier transform is given by:

$$\tilde{F}(f(x, y)) = \frac{1}{2\pi} \int_{-\infty}^{+\infty} \int_{-\infty}^{+\infty} f(x, y)e^{-i(k_x x + k_y y)} \, dx \, dy \tag{9}$$

where k_x and k_y are the spatial frequencies. In most cases, if one Fourier transforms the partial differential equation (8), then the simple ordinary differential equation in frequency space is obtained:

$$d\tilde{C}(k_x, k_y, t)/dt = -D(k_x^2 + k_y^2)\tilde{C}(k_x, k_y, t) \tag{10}$$

which has the solution:

$$\tilde{C}(k_x, k_y, t) = C_0(k_x, k_y, 0)e^{-(k_x^2 + k_y^2)Dt} \tag{11}$$

If one knows $C(k_x, k_y, t)$ one can return to $C(x, y, t)$ by taking the inverse Fourier transform

$$C(x, y, t) = \frac{1}{2\pi} \int_{-\infty}^{+\infty} \int_{-\infty}^{+\infty} \tilde{C}(k_x, k_y)e^{i(k_x x + k_y y)} \, dk_x \, dk_y \tag{12}$$

What is important to realize is that, whereas in the past one was limited analytically to solve such problems, image processing software and hardware can be used to determine Fourier transforms numerically. As a result, data can be analyzed using this simple form directly.

If we proceed from the general to the more specific, we can also employ the method of Koppel (1979, 1980) to simplify FRAP analysis. The normal mode of laser operation is where the laser profile can be described by a Gaussian in both the x and y directions. This results in the bleach profile $C(x, y, 0)$, being an exponential with a Gaussian argument; or, equivalently, as an infinite power series of Gaussians. Empirically it can be shown that only a few, and often only one, Gaussians are needed to describe $C(x, y, 0)$ to within experimental accuracy. The solution of the diffusion equation for a single Gaussian of width ω is of the form:

$$C(x, y, t) = \frac{C_0}{[1 + (8Dt/\omega^2)]} e^{-2(x^2+y^2)/\omega^2[1 + (8Dt/\omega^2)]} \tag{13}$$

in real space and,

$$\tilde{C}(k_x, k_y, t) = (C_o/4\omega^2)e^{-(k_x^2+k_y^2)[(\omega^2/8)+Dt]} \tag{14}$$

in frequency space. Thus, if $C(x, y, 0)$ can be described as a sum of N Gaussians, then the time-dependent solution will be given by

$$\tilde{C}(k_x, k_y, t) = \sum_{i=1}^{N} \frac{A_i}{4\omega_i^2} e^{-(k_x^2+k_y^2)[(\omega_i^2/8)+Dt]} \tag{15}$$

in frequency space, or by

$$C(x, y, t) = \sum_{i=1}^{N} \frac{A_i}{(1+8Dt/\omega_i^2)} e^{-2(x^2+y^2)/\omega_i^2[1+(Dt/\omega_i^2)]} \tag{16}$$

in real space. The Gaussian nature of each component of the initial distribution is preserved with time in both spaces.

In frequency space the amplitudes remain constant while the widths of the Gaussian decrease as

$$1/[(\omega_i^2/8) + Dt]^{1/2} \tag{17}$$

In real space the amplitudes decrease by

$$A_i/[1 + (8Dt/\omega_i^2)] \tag{18}$$

and the widths increase as

$$\omega_i[1 + (Dt/\omega_i^2)]^{1/2} \tag{19}$$

We can thus see two approaches to data analysis which simplify dealing with imaging FRAP data. The first is to Fourier transform the data and analyze in frequency space. The second is to empirically fit the data in either real or frequency space to a sum of Gaussians. The diffusion coefficients can then be determined from the time variations of the amplitudes or widths.

The expansion into Gaussians, along with the separability of Gaussians in Cartesian coordinates, enables one to extend this approach to the analysis of diffusional anisotropy, i.e., to measure diffusion that is different in the x and y directions.

4. RATIONALE FOR DEVELOPING VIDEO IMAGING FRET AND FRAP

As developed above, FRET is a solution technique which, as a result, involves signal averaging over many cells and all regions of the cell. Modification of conventional FRET measurements to the fluorescence-activated cell sorter enables one to do FRET measurements on a cell-by-cell basis but still averages the signal over the entire surface of a given cell. FRAP is a single-cell technique and, in fact, one can measure diffusibility on multiple regions of the same cell. In spot FRAP, the signal obtained is a weighted integral over a defined laser spot. As a result, all spatial and directional information is averaged out and lost. In extending these techniques by using a two-dimensional video detector rather than a photomultiplier tube, one can map transfer efficiencies over the cell surface, and spatial information lost in spot FRAP can be retained. As examples of the latter, we will show how video imaging FRAP can be used to measure diffusional anisotropy (i.e., whether the diffusibility is the same in all directions) and the rates of interregional diffusion (i.e., diffusional exchange between morphologically distinct regions of a cell). In addition, video imaging FRAP enables one to measure diffusion on moving cells and to readily separate the recovery due to diffusion from recovery due to membrane flows.

In both video imaging FRET and video imaging FRAP one must develop the ability to use a low-light-level video detector such as a SIT or cooled CCD camera as a quantitative device. In the next section we will consider the various types of corrections that one must make to accomplish this.

5. QUANTITATIVE IMAGING USING A VIDEO CAMERA

While it is not our goal here to consider all of the experimental problems of developing an imaging system in detail, it is useful to introduce the kinds of issues that must be resolved if one wishes to use a video detector as a sensing device that will provide meaningful spatial and quantitative information.

5.1. Detector Sensitivity and Signal-to-Noise Ratio

We must begin our discussion of using video cameras by introducing the concept of an image picture element, known as a pixel. Most video detectors are rectangular, and one can think of this rectangle as being divided like a piece of graph paper into proportionally smaller rectangular elements known as pixels. In fact, the detector of CCD cameras is a physical array of individual rectangular detectors. Typically, the image is divided into approximately a quarter of a million pixels. A signal, which would be spread over the entire surface of a photomultiplier tube, is now divided into a quarter of a million signals. Noise is stochastic in nature and therefore can be expected to vary as the square root of the signal. As a result, the uncertainty of a single pixel intensity is potentially 500 times noisier than that of the integrated signal. This considers only thermal noise and not other forms of noise intrinsic to the camera. This problem is further compounded by the rapid sampling rates of video cameras (30 frames per second) and for CCD cameras, by electronic readout noise.

Strategies for overcoming these problems include: cooling the detector to reduce thermal noise, and signal averaging both temporally and spatially. Methods of spatial signal averaging include a variety of convolution filters. The concept of a convolution will be discussed below.

As we will also discuss below, the number of pixels determines the spatial resolution of the system. One can readily see that in reducing the number of pixels to improve detector sensitivity, one sacrifices spatial resolution. We will also consider how one can use empty magnification to make the pixel size the determining factor in system spatial resolution. The downside of this is that while resolution increases with magnification, sensitivity decreases with the square of the magnification.

5.2. Gain Linearity and Dynamic Range

When using either a photomultiplier tube or a video detector, it is critical to operate the system on its linear range. That is to say, one wants the output signal to be linearly proportional to the input signal. If this is not the case, then one has to correct the data for these nonlinearities. While such corrections are tedious, the application of microcomputers to the laboratory makes them possible. Typically, a video camera has two gains which one must be concerned about. The first of these gains is a high voltage which governs the camera sensitivity. The second is an electronic gain which is a multiplication factor on the first gain. The complicated goal is to operate both of these gains in their optimal range. Operationally we have found, using a SIT camera, that when both these gains are operating optimally, one can obtain reasonable linearity over two orders of magnitude in light intensity (see Fig. 4).

5.3. Distortions and Inhomogeneities across the Detector Surface

In our discussions of video detectors so far we have assumed that the camera/microscope system is spatially homogeneous. In other words, we are assuming that the output image is a true representation of the object, that the shape of the object is undistorted, and that the relative intensities of different points are the same as the relative intensities of different points of the object. In reality, both the microscope and the camera introduce a number of distortions which must be corrected for.

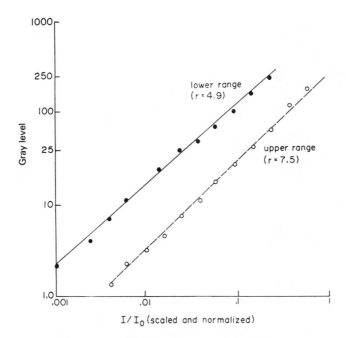

FIGURE 4. Gain response of typical SIT camera. Response (gray level) versus relative intensity for two incident light intensity values. The r values are rheostat settings in volts.

Figure 5 shows the image of a rectangular grid through a typical video microscope system. One observes that the grid image is "pincushioned." Effectively, the magnification increases as one moves from the center to the periphery of the field. Algorithms that correct for pincushioning are discussed in detail in texts on image processing (see, e.g., Castleman, 1979). One such useful algorithm is hyperbolic interpolation. One assumes that within a block, the distorted edges can be represented by four hyperbolas which share a common focal point at the center of the distorted square. These hyperbolas determine a mapping between real and distorted image space which can be applied to any element within the block. Such corrections are a nuisance to employ and often the problem is resolvable by confining one's self to the center of the image where no correction is necessary. It should be noted that pincushioning in addition to geometrically distorting the image causes a nonhomogeneous sensitivity of the detector to light over its surface. A pixel at the center of the image corresponds to a larger area of the object than does a pixel near the periphery. This nonhomogeneity falls out when one corrects for nonhomogeneous illumination and sensitivity as described below.

Further spatial inhomogeneities are introduced by the fact that one

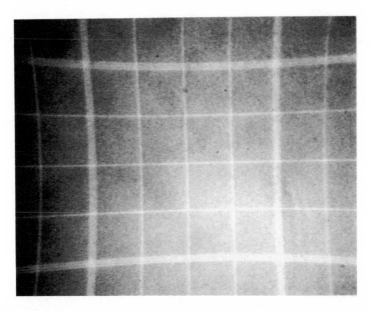

FIGURE 5. Image of a rectangular grid through the fluorescence microscope showing the geometric image distortion known as pincushioning. In addition to producing a spatial distortion, this phenomenon causes magnification to vary over the surface which, in turn, results in nonhomogeneous sensitivity to light.

cannot *a priori* expect that either the illumination of the system or the sensitivity of the detector will be the same over space. These nonhomogeneities are easily corrected for by dividing each image by a calibration image of a surface of homogeneous fluorophore concentration.

Even when one makes such a correction, one is left with the assumption that the linearity of the system is homogeneous across the field. If such is not the case, then this must be corrected digitally on a pixel-by-pixel basis.

A final aspect of the imaging system is to consider the point spread, or transfer function. Any optical system fuzzes out an image. If one were to look at the image of a point illumination, it would appear to be spread out over many pixels. This spread can best be mathematically represented as the point spread function. For simplicity, let us consider this problem in one dimension. The point spread function, $\rho(x' - x)$, convolves or spreads each point of the image. If $f(x)$ represents the true image and $i(x)$ represents the measured image, then

$$i(x) = \int_{-\infty}^{+\infty} f(x')\rho(x' - x)dx' \tag{20}$$

The trick is to solve this equation for $f(x)$. Once again, Fourier transforms can be used to transform Eq. (20) into frequency space. In frequency space, the convolution integral (20) reduces to the simple multiplication

$$\tilde{F}(i(x)) = \tilde{F}(f(x))\tilde{F}(\rho(x' - x)) \tag{21}$$

where, again, the symbol $\tilde{F}$ is taken to be the Fourier transform (Section 9.1). A derivation of this convolution theorem is given in Section 9.2. Here, the spatial frequencies k_x have units of 1/distance. One can then solve the equation for $\tilde{F}(f(x))$ and subsequently inverse Fourier transform to obtain $f(x)$. While this sounds complicated, it is in fact readily accomplished using numerical fast Fourier transform algorithms which are typically supplied with packaged image processing software.

More problematic is the fact that since the point spread function typically falls off very rapidly with distance, the low frequencies are difficult to determine numerically. It is usually necessary therefore to first fit the point spread function to an analytical function, and then to use the analytical Fourier transform of this function when applying Eq. (21). In most instances the two-dimensional point spread function can be fit to a two-dimensional Gaussian (Castleman, 1979). We see once again that the Gaussian is a mathematically convenient function by virtue of the fact that it remains Gaussian in frequency space. Figure 6, for instance, shows the point spread function of a SIT camera/microscope system.

One can really complicate the issue and ask whether the point spread function is the same all over the image. The full potential of point spread functions is revealed by experiments where the three-dimensional point spread function of a microscope imaging system is used to optically section microscopic objects (Fay *et al.*, 1986).

6. VIDEO IMAGING FRET (VIFRET)

The implementation of FRET to video microscope systems can be easily achieved providing that all sources of distortion are corrected as outlined in the previous section. The laboratory of F. Fay (Williams *et al.*, 1985) has pioneered the use of ratioing images at two wavelengths to produce maps of intracellular Ca^{2+} concentration. The technique of VIFRET has similar requirements, except that one typically wants to measure, rather than excite, at two wavelengths. Because FRET is sensitive to molecular distances of less than 100 Å, it can be used to study a variety of events in living cells. Therefore, appropriate choices of donor and acceptor fluorophores attached to macromolecules can yield important information about associations between different macromolecules within a cell.

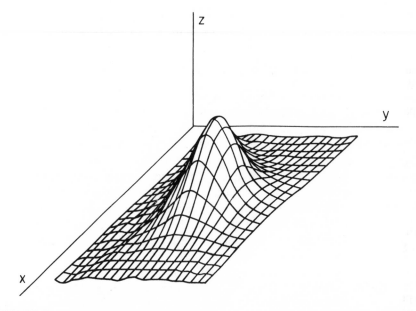

FIGURE 6. Three-dimensional representation of the point spread function for a video microscopy system. It can be shown that this function is well fitted to a Gaussian in both the *x* and *y* directions.

6.1. Modifying the Microscope for VIFRET

As discussed in Section 2.1, when the donor emission and the acceptor absorbance spectra overlap, energy transfer will occur, providing that donor and acceptor are sufficiently close to one another. Perhaps the most dramatic change which occurs upon FRET is a quenching of donor fluorescence with a concomitant appearance and increase of acceptor fluorescence ("sensitized emission") when only the donor fluorophore is directly excited. The FRET microscope requires two filter sets, one which excites and monitors donor fluorescence, and another which excites donor but monitors only acceptor fluorescence. It can also be useful to have a set which excites and monitors acceptor fluorescence.

The feasibility of this technique has been demonstrated by Uster and Pagano (1986), where energy transfer between either NBD or fluorescein as donor to rhodamine as acceptor was observed. Transfer is dramatically shown in Fig. 7. Two populations of large liposomes were prepared. One population contained NBD-labeled phosphatidylethanolamine (donor) and the second population also contained the NBD phosphatidylethanolamine and, in addition, contained a rhodamine labeled phosphatidylethanolamine.

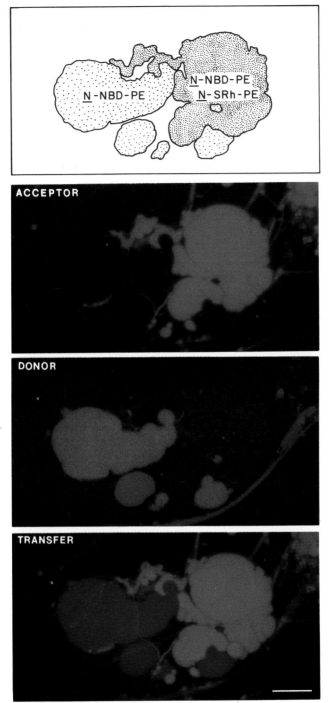

FIGURE 7. Energy transfer through a fluorescence microscope between NBD and Texas red-labeled phosphatidylethanolamine in model membranes. See text for details. Reprinted from Uster and Pagano (1986) with permission.

Figure 7 shows a field containing both populations of liposomes. The top panel shows the field under rhodamine illumination. Only the vesicle on the left shows up. The middle panel shows the field under NBD illumination. The right vesicles show up, the left one does not, because NBD fluorescence in that vesicle is quenched by the presence of acceptor rhodamine. The lower panel shows the field under conditions of exciting for NBD but monitoring for rhodamine. The right vesicles now show up brightly because of energy transfer. We also observe some fluorescence in the left vesicle. This is due to the tail of the NBD emission spectrum at these longer wavelengths. Uster and Pagano (1986) have also demonstrated that this technique can be used to localize fluorophore distributions in cells.

6.2. Quantifying FRET through the Microscope

The photographic technique discussed above is limited both by the long exposures, which can bleach and thus artificially quench samples, and by the intrinsic nonlinearity of photographic film. One can overcome these problems by using a video setup and correcting for any inherent nonlinearities as outlined in Section 5. In theory, quantifying VIFRET is a simple matter of quantifying donor and acceptor fluorescence intensities with the video camera. However, a number of controls must be performed in order to verify that nonradiative FRET is occurring. A typical setup of fluorescence filters should include: a donor excitation and emission cube as well as a cube that excites the donor but only looks at acceptor emission fluorescence (the transfer cube). With these two cubes one can monitor the donor fluorescence in the absence of acceptor, I_d, acceptor fluorescence in the absence of donor when exciting at donor excitation wavelengths, I_a, donor fluorescence in the presence of acceptor, $I_{d,a}$, and sensitized acceptor fluorescence, $I_{a,d}$, in the presence of donor.

In order to accurately calculate the transfer efficiency of a given donor–acceptor pair, one must make measurements of the four intensities given above. These values can be stored within a computer and used for calculations of transfer efficiency, over selected areas, at a later time. In addition, corrections must be made for bleaching rates of the fluorophores and, in the case of cell surface molecules, background levels of fluorescence must be subtracted from the measured values on the cell surface.

To quantify transfer efficiency, E_t, on the surface of membranes, an area of interest over the cell is first determined. This area is used for all measurements of intensity outlined above. Since there are a number of phenomena that can lead to either donor quenching or acceptor enhancement, the transfer efficiency must be calculated *both* from the quenching of donor fluorescence and from the enhancement of acceptor fluorescence:

$$E_t(d) = (I_d - I_{d,a})/I_d \tag{22}$$

and

$$E_t(a) = I_{a,d}/I_{a,d \text{ max}} \tag{23}$$

where the intensities, I_d, I_a, $I_{d,a}$, and $I_{a,d}$, have been corrected for background and bleaching over a time t and $I_{d,a \text{ max}}$ is the extrapolated sensitized acceptor emission at infinite acceptor concentration. If FRET is the only process working, then the transfer efficiency calculated from the donor quenching, $E_t(d)$, will be equal to the transfer efficiency calculated from acceptor enhancement, $E_t(a)$. Once calculated, the transfer efficiencies can be used along with Eq. (1) to calculate the intermolecular distance between the donor and acceptor. *Thus, using VIFRET, the fluorescence microscope can be used to obtain spatial information about two orders of magnitude below the optical resolution of the microscope.*

7. VIDEO IMAGING FRAP (VIFRAP)

In contrast to VIFRET, VIFRAP has been used both to observe and to quantify fluorescence events over small regions of a cell. As stated in previous sections, the advantage of VIFRAP over conventional spot photobleaching is that spatial information is retained over the entire time course of an experiment. This is useful for measuring events within cells that might otherwise be averaged out using conventional FRAP techniques (e.g., interregional diffusion, anisotropic diffusion, macromolecular assembly).

7.1. Observing Cellular Dynamics

The technique of FRAP has been used to study a variety of phenomena both within the cell and on the cell surface (for reviews see Peters, 1981; Wolf, 1989). FRAP within cells is not trivial and is potentially laden with artifacts (Vigers *et al.*, 1988; Simon *et al.*, 1988). In general, one must keep the bleach duration short so that potential artifacts such as local heating and photodamage resulting in free radical formation and subsequent polymerization of macromolecules are avoided. In addition, if a deep bleach is needed, the specific bleach profile in three dimensions must be considered if data are to be analyzed correctly since the bleach depth may be significantly greater than the focal length of the objective and recovery may have both radial and transverse components.

One dramatic application of VIFRAP within cells was in observing actin treadmilling within cells as described by Wang (1985). In these experiments, gerbil fibroma cells were microinjected with actin that had been fluorescently labeled with iodoacetamidotetramethylrhodamine. The amount of labeled actin was estimated to be between 5 and 10% of the total endogenous actin pool. In addition, injection and observation occurred only in the lamellipodium ("leading edge") of the fibroblasts where the cell is particularly thin and the actin is highly polarized with the barbed end of the actin being attached to the membrane. Various investigators had suggested that actin treadmilling occurred at different rates at the "barbed" and "pointed" ends of the actin filament. To test this hypothesis, Wang photobleached actin microspikes at the surface of the membrane and viewed the recovery using enhanced video microscopy techniques. In all experiments, recovery of the fluorescently labeled actin began at the membrane and extended toward the center of the cell (Fig. 8). Wang also found that the rate of recovery toward the center of the cell was approximately 0.8 μm/sec. From these data, he concluded that the steady-state incorporation of actin subunits occurred predominately at the "barbed" end, supporting the hypothesis that this was a faster polymerization process than that at the "pointed" end. This also suggested that, in nonmuscle cells, treadmilling of actin may represent a mechanism for cellular expansion in the absence of myosin.

Another example of viewing a dynamic process is the diffusion of lipids through different morphological regions on the sperm cell surface as described by Wolf and Voglmayr (1984). In these experiments, cells were labeled with the fluorescent lipid analogue, 1,1'-dihexadecyl 3,3,3',3'-tetramethylindocarbocyanine perchlorate (C_{16} diI). Carbocyanine dyes have been used extensively as membrane probes, especially for the measurement of lateral diffusion (for review see Peters, 1981). Using conventional spot photobleaching techniques, Wolf and Voglmayr (1984) found that the three major morphologically distinct regions on the sperm surface (the head, midpiece, and tail) had distinct diffusion characteristics. In addition, these investigators found, using intensified video microscopy, that when the fluorescence from one morphological region of the cell was completely bleached, recovery occurred over a few minutes. Figure 9 shows a ram sperm head that has been completely bleached with subsequent recovery occurring over 5 min. These results show that, although the different morphological regions of the sperm surface had unique diffusion characteristics, free exchange of C_{16} diI could occur between these different regions. That is, the difference in diffusion characteristics did not, in itself, reflect a barrier to diffusion between these regions. The two examples described above are largely qualitative in the information that they provide. In the next section, we will describe how the processes of interregional diffusion and anisotropic diffusion can be further quantified using VIFRAP. For a further discussion of VIFRAP the reader is referred to Kapitza et al. (1985).

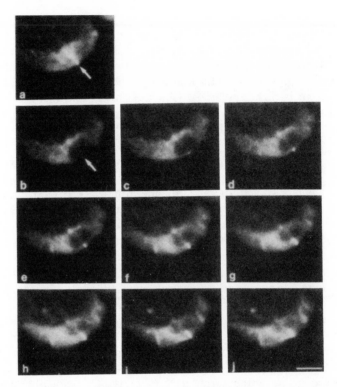

FIGURE 8. VIFRAP experiment showing treadmilling of rhodamine-labeled actin filaments in gerbil fibroma cells. See text for details. The cell before photobleaching is shown in a. A microspike (arrow) and its surrounding area are photobleached at $t = 0$. Images were then recorded at $t = 35$ sec (b), 63 sec (c), 95 sec (d), 123 sec (e), 160 sec (f), 188 sec (g), 217 sec (h), 245 sec (i), and 275 sec (j). The recovery begins at the membrane and moves toward the center of the cell. Bar = 5 μm; $\times$1570. Figure and legend from Wang (1985) with permission.

7.2. Quantitation of Diffusion Events

As indicated above, the problem of producing quantitative data from an image is twofold. First, one must correct for intrinsic nonlinearities, inhomogeneities, and backgrounds. Second, algorithms that enable one to deal with large amounts of information and inherant noise must be developed. As an example, let us consider how one would measure diffusional anisotropy within a membrane. Figure 10 is an example of such an experiment. A 3T3 cell was labeled at the cell surface with the lipid probe C_{16} diI. A spot of approximately 10 μm was bleached and the intensity sampled in the horizon-

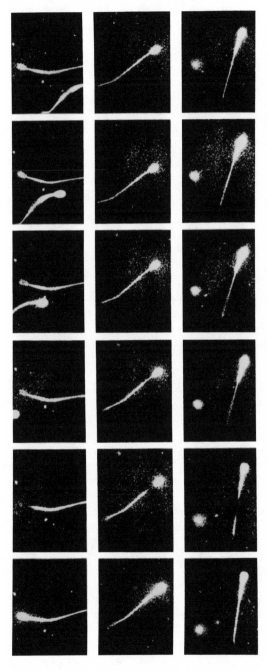

FIGURE 9. VIFRAP experiment demonstrating interregional diffusion of the fluorescent lipid analogue, C_{16} dil, between the major morphological regions of ram spermatozoa. See text for details. The first column in each series is prebleach. The second column is immediately postbleach, and each subsequent column is 60 sec after the previous one. ×500. From Wolf and Voglmayr (1984) with permission.

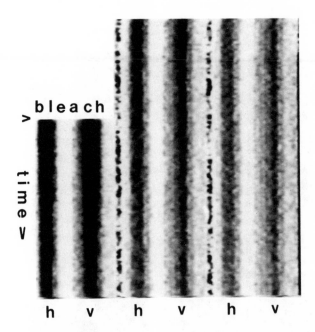

FIGURE 10. VIFRAP experiment to study anisotropic diffusion of lipid in a 3T3 cell plasma membrane. See text for details.

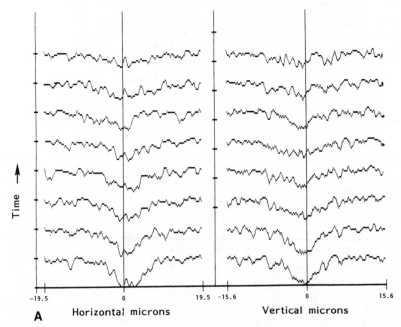

FIGURE 11. VIFRAP experiment similar to that of Fig. 10 to measure rhodamine-labeled BSA diffusion in a 90% glycerol solution. See text for details. (A) Data corrected for background, inhomogeneities, and nonlinearities and normalized to pre-bleach intensity values. (B) Data in A have been low pass filtered to reduce noise. (C) The data were then fit to a single Gaussian.

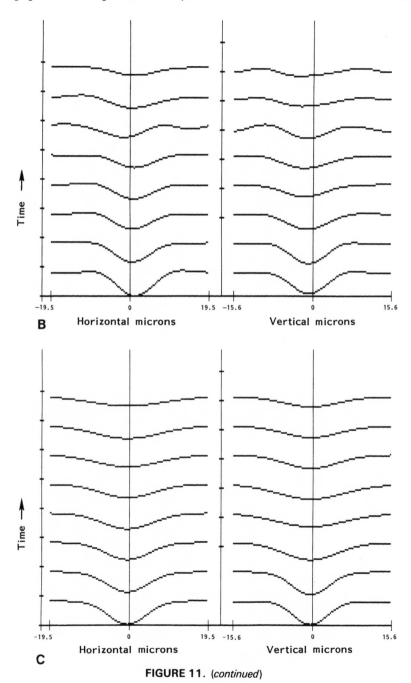

B   Horizontal microns   Vertical microns

C   Horizontal microns   Vertical microns

FIGURE 11. (*continued*)

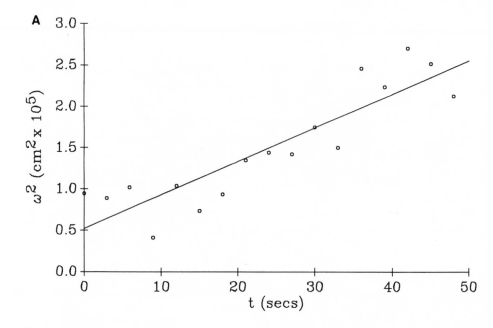

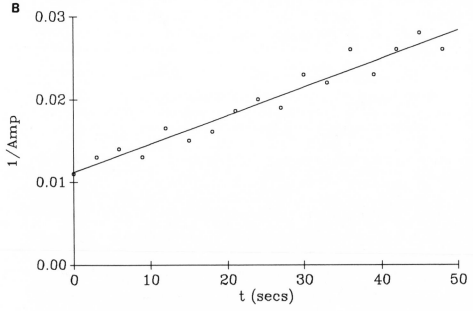

FIGURE 12. Data of Fig. 11C displayed as (A) ω^2 and (B) $1/A$ as a function of time to show how D can be calculated. See text for details.

tal and vertical directions at 0.17-sec intervals. These horizontal and vertical intensity profiles were stored side by side sequentially in the frame buffer so that, as shown in Fig. 10, time increased from top to bottom and from left to right. The data in Fig. 10 represent an image that has been corrected for background, inhomogeneities, and nonlinearities over 1 min. To show how such images can be processed, consider Fig. 11 which shows intensity versus position profiles for a similar experiment measuring rhodamine-BSA diffusion in a 90% glycerol solution at room temperature. A flat, 50-μm-pathlength capillary tube was used to simulate a two-dimensional situation and a 25× objective employed. Figure 11A shows data corrected for background, inhomogeneities, and nonlinearities and normalized to prebleach intensity values. Here we show only curves at 6-sec intervals. Tick marks in the vertical direction indicate the baseline for the corresponding curve. Figure 11B shows the same data after using a low pass filter to reduce high-frequency noise. The data in Fig. 11A were also fitted to a single Gaussian demonstrating that this is a reasonable approximation to the filtered data set (Fig. 11C). Note that recovery manifests itself as both a reduction in amplitude and a filling and broadening of the width of the Gaussian. As predicted by Eq. (18) and (19), if one plots 1/amplitude and $1/\omega^2$ as a function of time, one should obtain linear plots with slopes of $8D/A_o\omega_o^2$ and $8D$, respectively. This is shown in Fig. 12a and 12b for the data of Fig. 11. Using Eq. (18) we obtained $D = 1.9 \times 10^{-9}$ cm²/sec and using Eq. (19) $D = 5.1 \times 10^{-9}$ cm²/sec. This can be compared with the theoretical value of 2.6×10^{-9} cm²/sec derived using the Stokes–Einstein equation.

We wish to point out that this analysis represents preliminary studies. One can expect that as recovery progresses, systematic errors will be introduced because as the amplitude is decreasing and as ω increases, one is progressively sampling a smaller fraction of the Gaussian. It therefore becomes important to characterize the requirements of the fitting routines to correctly determine the recovery over long times.

8. CONCLUSIONS

In this chapter we have shown that digitally enhanced video fluorescence microscopy can be used to extract important spatial and temporal information from cells. Physical techniques that have been previously limited to averaging spatial information over relatively large distances can now be used to measure important parameters such as macromolecular diffusion coefficients and intermolecular distances up to 100 Å. Specifically, the development of VIFRAP allows one to not only measure diffusion coefficients and percent recoveries, but also to answer questions about the nature of diffusion through an inhomogeneous milieu and across boundaries. Likewise, the development of VIFRET allows one to determine distances between molecules of putative physiological importance within different cellular domains. Application of these techniques using digitally enhanced video microscopy, as well as others yet to be developed, will undoubtedly alter our view of cellular microarchitecture and function.

9. APPENDIX: CALCULATIONS IN RECIPROCAL SPACE

In analyzing closely spaced events, it is often more convenient to perform all mathematical manipulations in reciprocal, or frequency, space and then transform the final result back into the original coordinate system. The reader is perhaps most familiar with the concept of reciprocal space in the time domain where the reciprocal transformation places one in the frequency (f) domain. Likewise, in configuration space (i.e., an x, y, z coordinate system) an orthogonal transformation takes one into spatial frequency space (k_x, k_y, k_z). As will be shown, many operations used in digital enhanced video microscopy, such as convolutions, are trivial in reciprocal space. Furthermore, these principles can be used to solve complicated equations, such as the two-dimensional diffusion equation, which are often impossible to analyze without these transformations.

9.1. Fourier Transforms

The Fourier transform defines the transformation of a set of data, in one coordinate system, into reciprocal space. By performing this transformation, one is able to quantify the effects of digitizing systems which would not be possible in configuration space. In one dimension, the Fourier transform is defined as:

$$\mathcal{F}(k_x) \equiv \tilde{F}(f(x)) = \frac{1}{\sqrt{2\pi}} \int_{-\infty}^{+\infty} f(x)e^{-ik_x x}\, dx \qquad (A1)$$

where $f(x)$ is the function to be transformed, $\mathcal{F}(k_x)$ is the transform itself, and i is the complex number, $\sqrt{-1}$. Simply put, the Fourier transform represents an integral transformation which takes one complex function of n real variables into another complex function of n real variables. Conversely, the inverse Fourier transform is defined as:

$$\tilde{F}^{-1}(\mathcal{F}(k_x)) = \frac{1}{\sqrt{2\pi}} \int_{-\infty}^{+\infty} \mathcal{F}(k_x)e^{ik_x x}\, dk_x = f(x) \qquad (A2)$$

and allows one to readily return to the original coordinate system. The functions $f(x)$ and $\mathcal{F}(k_x)$ represent a Fourier transform pair. Note that these two transformations differ only by a minus sign in the exponential.

The fact that the transformation is reciprocal is a consequence of Fourier's integral theorem:

$$f(x) = \int_{-\infty}^{+\infty} \left[\int_{-\infty}^{+\infty} f(x) e^{-ik_x x} \, dx \right] e^{ik_x x} \, dk_x \tag{A3}$$

which means that:

$$\tilde{F}(f(x)) = \mathcal{F}(k_x) \Rightarrow \tilde{F}^{-1}(\mathcal{F}(k_x)) = f(x) \tag{A4}$$

Tables of Fourier transforms (and inverse Fourier transforms) of frequently encountered functions are available in many standard books of math tables. Of particular relevance for our discussion is the Fourier transform of a constant, a:

$$\tilde{F}(a) = \frac{a}{\sqrt{2\pi}} \int_{-\infty}^{+\infty} e^{-ik_x x} \, dx = a\delta(k_x) \tag{A5}$$

where $\delta(k_x)$ is the delta function (an impulse). Another function of interest, especially in analyzing VIFRAP data, is the Fourier transform of a Gaussian e^{-ax^2}:

$$\tilde{F}(e^{-ax^2}) = \frac{1}{\sqrt{2\pi}} \int_{-\infty}^{+\infty} e^{-ax^2 - ik_x x} \, dx = \frac{1}{\sqrt{2a}} e^{-k_x^2/4a} \tag{A6}$$

That is, the Fourier transform of a Gaussian in configuration space is simply a Gaussian in reciprocal space.

Fourier transforms have many interesting and useful properties, especially when dealing with large arrays of data, such as images. For a review of these properties, the reader is referred to a number of excellent texts (Castleman, 1979; Makovski, 1983). Fourier transforms lend themselves beautifully to analysis by computer. The discrete Fourier transform (DFT) and the fast Fourier transform (FFT) allow one to rapidly analyze a large amount of data in reciprocal space that would be prohibitive in configuration space. At this time, an image representing 512 pixels $\times$ 512 pixels $\times$ 8 bits can be fully transformed within 10 sec on a standard microcomputer.

9.2. The Convolution Theorem

In our discussion of the point spread function, we encountered the concept of convolution. In general, a convolution of two functions, $f(x)$ and $g(x)$, is given by:

$$\int_{-\infty}^{+\infty} f(x')g(x'-x)\,dx' \tag{A7}$$

the shorthand notation for this operation is simply:

$$f(x)*g(x) \tag{A8}$$

The mathematical process of convolving (and "deconvolving") an image is complicated in configuration space. Beyond the point spread function, convolutions are useful in such image processes as filtering and edge enhancement. A very useful property of Fourier transforms is encompassed in the so-called Shift theorem, which describes the effect that moving the origin of a function has upon its transform. Simply, the effect of shifting the origin, by some distance b, adds another complex exponential into its transform. The Shift theorem states:

$$\tilde{F}(f(x-b)) = \frac{e^{-ik_x b}}{\sqrt{2\pi}} \int_{-\infty}^{+\infty} f(u)e^{-ik_x u}\,du = e^{-ik_x b}\mathscr{F}(k_x) \tag{A9}$$

where $u = x - b$ and $du = dx$.

The Fourier transform of the convolution (A8) can thus be written as:

$$\tilde{F}(f(x)*g(x)) = \frac{1}{2\pi} \int_{-\infty}^{+\infty} \int_{-\infty}^{+\infty} f(u)g(x-u)e^{-ik_x x}\,dx\,du \tag{A10}$$

which, upon collecting terms, can be written as:

$$\tilde{F}(f(x)*g(x)) = \frac{1}{2\pi} \int_{-\infty}^{+\infty} f(u) \int_{-\infty}^{+\infty} g(x-u)e^{-ik_x x}\,dx\,du \tag{A11}$$

Applying the Shift theorem to Eq. (A11) yields:

$$\tilde{F}(f(x)*g(x)) = \frac{\mathscr{G}(k_x)}{\sqrt{2\pi}} \int_{-\infty}^{+\infty} f(u)e^{-ik_x u}\,du \tag{A12}$$

where

$$\mathscr{G}(k_x) \equiv \tilde{F}(g(x))$$

or

$$\tilde{F}(f(x)*g(x)) = \mathscr{G}(k_x)\mathscr{F}(k_x) \tag{A13}$$

and

$$\tilde{F}^{-1}(\mathscr{G}(k_x)\mathscr{F}(k_x)) = f(x)*g(x) \tag{A14}$$

Hence, the complicated problem of convolution reduces to simple multiplication in reciprocal space, thereby greatly simplifying the degree of convolution that must be performed.

9.3. Solution of the Diffusion Equation in Fourier Space

To solve Eq. (8) we will use the Fourier transform. Recall the two useful Fourier transforms shown in Section 9.1.

- The Fourier transform of a constant a:

$$\tilde{F}(a) = a\delta(k_x) \tag{A5}$$

- The Fourier transform of a Gaussian e^{-ax^2}:

$$\tilde{F}(e^{-ax^2}) = \frac{1}{\sqrt{2a}} e^{-k_x^2/4a} \tag{A6}$$

Now if one Fourier transforms Eq. (8), one obtains

$$\frac{\partial \tilde{C}(k_x, k_y, t)}{\partial t} = -D(k_x^2 + k_y^2)\tilde{C}(k_x, k_y, t) \tag{A15}$$

This is an ordinary differential equation and has the solution

$$\tilde{C}(k_x, k_y, t) = \tilde{C}(k_x, k_y, 0)e^{-(k_x^2+k_y^2)Dt} \tag{A16}$$

The idea now is to determine the Fourier transform, $\mathscr{F}\{C(x, y, 0)\}$, of the initial condition, $C(x, y, 0)$, and then to inverse Fourier transform (A16) to obtain $C(x, y, t)$.

Now if one is bleaching with a Gaussian laser beam, the bleach profile immediately postbleach looks like a Gaussian subtracted from the uniform prebleach concentration C_o. Let us assume that such is the case so that we assume

$$C(x, y, 0) = C_o - Ae^{-2(x^2+y^2)/\omega^2} \tag{A17}$$

where ω is the beam radius. Gaussians have the convenient property that they can be separated in Cartesian coordinates into the product of a function of x and a function of y. If this is true for the initial condition, then it is also true for all later times.

The Fourier transform of (A17) is therefore

$$\tilde{C}(k_x, k_y, 0) = C_o\delta(k_x, k_y) - \frac{A\omega^2}{4} e^{-(k_x^2+k_y^2)(\omega^2/8)} \tag{A18}$$

Putting (A18) into (A16) we obtain

$$\tilde{C}(k_x, k_y, t) = C_o\delta(x, y)e^{-(k_x^2+k_y^2)Dt} - A(\omega^2/4)e^{-(k_x^2+k_y^2)[(\omega^2/8)+Dt]} \tag{A19}$$

To obtain the real space solution we inverse transform (A19) to obtain

$$C(x, y, t) = C_o - \frac{A}{[1 + (8Dt/\omega^2)]} e^{-2(x^2+y^2)/\omega^2[1+(8Dt/\omega^2)]} \tag{A20}$$

Now let us suppose that a sum of N Gaussians is required to describe $C(x, y, 0)$, namely

$$C(x, y, 0) = C_0 - \sum_{i=1}^{N} A_i e^{-2(x^2+y^2)/\omega_i^2} \quad \text{(A21)}$$

In this case (A19) becomes

$$\tilde{C}(k_x, k_y, t) = C_0 \delta(k_x, k_y) e^{-2(k_x^2+k_y^2)Dt}$$

$$- \sum_{i=1}^{N} \frac{A_i \omega_i^2}{4} e^{-2(k_x^2+k_y^2)/[(\omega_i^2/8)+Dt]} \quad \text{(A22)}$$

and (A20) becomes

$$C(x, y, t) = C_0 - \sum_{i=1}^{N} \frac{A_i}{[1 + (8Dt/\omega_i^2)]} e^{-2(x^2+y^2)/\omega_i^2[1+(8Dt/\omega_i^2)]} \quad \text{(A23)}$$

If one assumes that the laser beam is Gaussian with intensity profile

$$I = (I_0/a)e^{-2(x^2+y^2)/\omega^2} \quad \text{(A24)}$$

where a is an attenuation factor, then the intensity of fluorescence obtained in a spot FRAP experiment is given by:

$$I_F(t) = \frac{q}{a} \int_{-\infty}^{+\infty} \int_{-\infty}^{+\infty} C(x, y, t) I_0 e^{-2(x^2+y^2)/\omega^2} \, dx \, dy \quad \text{(A25)}$$

where q is the product of efficiencies for light absorption, detection, and emission. Substituting (A23) into (A25) yields:

$$I_F(t) = \frac{qI_0}{a} \left(\int_{-\infty}^{+\infty} \int_{-\infty}^{+\infty} C_0 e^{-2(x^2+y^2)/\omega^2} \, dx \, dy - \sum_{i=1}^{N} \frac{A_i}{[1 + (8Dt/\omega_i^2)]} \right.$$

$$\left. \times \int_{-\infty}^{+\infty} \int_{-\infty}^{+\infty} e^{\{-2(x^2+y^2)/\omega_i^2[1+(8Dt/\omega_i^2)]\}-2(x^2+y^2)/\omega^2} \, dx \, dy \right) \quad \text{(A26)}$$

After some algebra, these integrals and sums reduce to:

$$I_F(t) = \frac{qI_0\pi\omega^2}{2a} \left(C_0 - \sum_{i=1}^{N} \frac{A_i}{1 + (\omega^2/\omega_i^2)[1 + (2t/\tau_D)]} \right) \quad \text{(A27)}$$

where τ_D is defined as $\omega^2/4D$.

So far we have assumed that the bleach profile can be expressed in terms of a sum of Gaussians. This is essentially the approach of Koppel (1980) and is a very useful way to deal with the data especially if one intends to do imaging FRAP. In spot FRAP it is usually assumed that the bleaching process is first order (Axelrod *et al.*, 1976). Thus, if $C(x, y, t)$ is the fluorophore concentration at a position (x, y) at time t, $I(x, y)$ is the intensity profile of the bleaching light, and α is the bleaching rate constant, then

$$dC(x, y, t)/dt = -\alpha I(x, y)C(x, y, t) \tag{A28}$$

If one bleaches for time T, the fluorophore concentration becomes

$$C(x, y, 0) = C_o e^{-\alpha T I(x, y)} \tag{A29}$$

where C_o is the prebleach fluorophore concentration and where we have defined $t = 0$ to be at the end of the bleach. It is also useful to define a parameter κ which is a measure of the degree of bleaching:

$$\kappa \equiv \alpha T I_o \tag{A30}$$

For a Gaussian beam profile, therefore:

$$C(x, y, 0) = C_o e^{-\kappa e^{-2(x^2+y^2)/\omega^2}} \tag{A31}$$

The imposing form may be made palatable by expanding as a power series of $\exp[-2(x^2 + y^2)/\omega^2)]$.

Thus,

$$C(x, y, 0) = C_o \sum_{n=0}^{\infty} \frac{(-\kappa)^n}{n!} e^{-2n(x^2+y^2)/\omega^2} \tag{A32}$$

or, expanding out the zeroth term:

$$C(x, y, 0) = C_o + C_o \sum_{n=1}^{\infty} \frac{(-\kappa)^n}{n!} e^{-2n(x^2+y^2)/\omega^2} \tag{A33}$$

Thus, the initial concentration is given by C_o plus an infinite sum of Gaussians and we can use (A27) where $A_i = -C_0(-\kappa)^n/n!$ and $\omega_i^2 = \omega^2/n$.

We thus obtain the solution of Axelrod *et al.* (1976), namely

$$I_F(t) = \frac{q I_o \pi \omega^2}{2a} C_o \sum_{n=0}^{\infty} \frac{(-\kappa)^n}{n!} \frac{1}{1 + n[(1 + 2t/\tau_D)]} \tag{A34}$$

ACKNOWLEDGMENTS. The authors are grateful to Ms. Christine McKinnon and Ms. Kristine Bocian for technical assistance in the preparation of the manuscript. We also wish to thank Drs. Richard Pagano and Yu-Li Wang for graciously providing us with photographs from their work. This work was supported in part by NIH grants HD17377 (D.E.W.), HD23294 (D.E.W.), and HD07312 (R.A.C.). Additional funding was provided by the Whittaker Foundation (D.E.W.) and from the A. W. Mellon Foundation to the Worcester Foundation. Portions of this chapter represent work done in partial completion of the degree of Master of Science in Biomedical Engineering

(R.M.M.) at the Worcester Polytechnic Institute and we wish to thank Drs. Robert Peura and Steven Moore for their assistance.

10. REFERENCES

Axelrod, D., Koppel, D. E., Schlessinger, J., Elson, E., and Webb, W. W., 1976, *Biophys. J.* **16**:1055–1069.

Baird, B., and Holowka, D., 1988, in *Spectroscopic Membrane Probes* (L. M. Loew, ed.), CRC Press, Boca Raton, pp. 93–116.

Cardullo, R. A., Agrawal, S., Flores, C., Zamecnik, P., and Wolf, D. E., 1988, *Proc. Natl. Acad. Sci. USA* **85**:8790–8794.

Castleman, K. R., 1979, *Digital Image Processing,* Prentice–Hall, Englewood Cliffs, N.J.

Dewey, T. G., and Hammes, G. G., 1980, *Biophys. J.* **32**:1023–1035.

Fay, F. S., Fogarty, K. E., and Coggins, J. M., 1986, in *Optical Methods in Cell Physiology* (P. DeWeer and P. Salzburg, eds.), Wiley, New York, pp. 51–62.

Fung, B. K. K., and Stryer, L., 1978, *Biochemistry* **17**:5241–5248.

Inoue, S., 1986, *Video Microscopy,* Plenum Press, New York.

Kapitza, H. G., McGregor, G., and Jacobson, K. A., 1985, *Proc. Natl. Acad. Sci. USA* **82**:4122–4126.

Kleinfeld, A., 1988, in *Spectroscopic Membrane Probes* (L. M. Loew, ed.), CRC Press, Boca Raton, pp. 63–92.

Koppel, D. E., 1979, *Biophys. J.* **28**:281–292.

Koppel, D. E., 1980, *Biophys. J.* **30**:187–192.

Koppel, D. E., Fleming, P. J., and Strittmatter, P., 1979, *Biochemistry* **18**:5450–5464.

Makovski, A., 1983, *Medical Imaging,* Prentice–Hall, Englewood Cliffs, N.J.

Peters, R., 1981, *Cell Biol. Int. Rep.* **5**:733–760.

Simon, J. R., Gough, A., Urbank, E., Wang, F., Lanni, F., Ware, B. R., and Taylor, D. L., 1988, *Biophys. J.* **54**:801–816.

Snyder, B., and Friere, E., 1982, *Biophys. J.* **40**:137–148.

Taylor, D. L., and Wang, Y.-L. (eds.), 1989, *Quantitative Fluorescence Microscopy: Imaging and Spectroscopy,* Vol. 30 and 31, Academic Press, New York.

Uster, P. S., and Pagano, R. E., 1986, *J. Cell Biol.* **103**:1221–1234.

Vigers, G. P. A., Coue, M., and McIntosh, J. R., 1988, *J. Cell Biol.* **107**:1011–1024.

Wang, Y.-L., 1985, *J. Cell Biol.* **101**:597–602.

Williams, D. A., Fogarty, K. E., Tsien, R. Y., and Fay, F. S., 1985, *Nature* **318**:558–561.

Wolf, D. E., 1989, in *Quantitative Fluorescence Microscopy:* Methods in Cell Biology, Vol. 30 (D. L. Taylor and Y.-L. Wang, eds.), Academic Press, New York, pp. 271–306.

Wolf, D. E., and Voglmayr, J. F., 1984, *J. Cell Biol.* **98**:1678–1684.

Chapter 9

The Dynamic Parameter

Fluorescence Photobleaching as a Tool
to Dissect Space in Biological Systems

**Melvin Schindler, Paramjit K. Gharyal,
and Lian-Wei Jiang**

1. INTRODUCTION

The viability of organisms is dependent on the controlled flow of information and metabolic/synthetic precursors between cellular compartments.
Such processes are elaborated upon as a hierarchy of interdependence established between cells and tissues. Through the ebb and flow of signaling and
metabolic molecules, dynamic linkages may be maintained between cells for
the coordination, synchronization, and initiation of cellular cycles (Fig. 1).
In this manner, organismal response to the environment may be viewed as
the result of a linked web of dissipative molecular gradients across biological
membranes that initiate and transmit environmental information and cellular status. Integration of these gradients over large numbers of cells and
tissues collectively leads to spatial and/or temporal responses. The biological
structures that serve as controllable elements for transmembrane molecular
flow are generally classified as channels or pores that serve either as passive
transport routes for low-molecular-weight molecules (Loewenstein, 1979;
Nikaido and Nakae, 1979; Gunning and Overall, 1983) or as ion pumps or
transporters requiring some type of coupled gradient dissipation or energy

MELVIN SCHINDLER, PARAMJIT K. GHARYAL, and LIAN-WEI JIANG • Department of Biochemistry, Michigan State University, East Lansing, Michigan 48824.

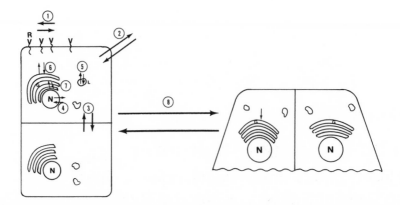

FIGURE 1. Dynamic linkages between organelles and cells. Chemical gradients are utilized to transmit information between the cell and the environment. The pathways involved in this transmission system are: lateral mobility of membrane receptors (1); transplasma membrane transport through channels, pores, and transporters (2); homotypic intercellular communication through gap junctions or plasmodesmata (3); nucleocytoplasmic transport (4); translysosomal or vacuolar membrane transport of H⁺ and ions (5); Golgi-mediated processing, secretion, and recycling (6); Golgi transport of newly synthesized proteins (*cis*-medial-*trans*) (7); heterotypic intercellular communication (8). R, N, and G represent membrane receptors, the nucleus, and Golgi, respectively.

source for molecular transposition (Mitchell, 1979; Noma, 1983; Reuter *et al.*, 1983). In most instances, the control of these channels is mediated by ligand-specific receptors that couple to the channels under activating conditions, initiating a cascade of enzymatic changes resulting in a modification of channel transport properties (Koshland, 1981; Bean *et al.*, 1983; Hondeghem and Katzung, 1984). In other cases, transport channels and receptors are intimately linked, forming a common structure, as in the case of the nicotinic acid receptor/channel (Conti-Tronconi and Raftery, 1982).

During the past several years, our laboratory has attempted to explore mechanisms and patterns of information flow at the plasma membrane (Koppel *et al.*, 1981; Metcalf *et al.*, 1983), cellular compartmental level (Schindler *et al.*, 1985; Jiang and Schindler, 1986), and between cells in tissue culture (Wade *et al.*, 1986; Baron-Epel *et al.*, 1988a). These investigations have required the development of sophisticated fluorescence-based analytical instrumentation to pursue subcellular transport measurements in viable single cells (Schindler *et al.*, 1989). This chapter will attempt to outline some of the particular biological questions in molecular information flow that we have pursued while also providing some description of the fluorescence-based instrumental approaches. The technological flexibility inherent in these approaches suggests that these methods may be utilized to investigate a broad range of communication questions at either the single or multiple cell level.

2. FLUORESCENCE REDISTRIBUTION AFTER PHOTOBLEACHING (FRAP)—THE OPTICAL MACHINE GUN

To measure molecular transport across membranes or cell walls, a means of creating and monitoring a spatial molecular gradient must be developed. It would be particularly advantageous if such a gradient could be repetitively established at will, giving the experimenter the flexibility to perform multiple analyses on the same biological material. Over the past years, the most successful technique for creating measurable dynamic gradients has been the method of FRAP (Koppel, 1979; Koppel *et al.*, 1981; Metcalf *et al.*, 1983; Schindler *et al.*, 1985; Kapitza and Jacobson, 1986; Jiang and Schindler, 1986; Wade *et al.*, 1986; Baron-Epel *et al.*, 1988a). The method in broad outline requires that fluorescent probe reporter molecules be introduced into either cell membranes or cytoplasm. These tagged molecules diffuse in most instances to create a homogeneous distribution of fluorescence as observed by excitation with the appropriate wavelength of light. A gradient of fluorescence may be established by photobleaching either a cellular area or volume of fluorescence (Koppel *et al.*, 1981; Metcalf *et al.*, 1983; Schindler *et al.*, 1985; Jiang and Schindler, 1986; Wade *et al.*, 1986; Baron-Epel *et al.*, 1988a). The net result of photobleaching is the creation of a nonfluorescent patch of probe molecules. The movement of molecules by diffusion, directed flow, or pump-mediated processes results in the reestablishment of fluorescence in the bleached area (membrane) or volume (cytoplasm). Appropriate mathematical analysis of fluorescence redistribution is utilized to calculate diffusion coefficients or transport fluxes (Koppel, 1979; Koppel *et al.*, 1981; Metcalf *et al.*, 1983; Schindler *et al.*, 1985; Jiang and Schindler, 1986; Kapitza and Jacobson, 1986; Wade *et al.*, 1986; Baron-Epel *et al.*, 1988a). A cartoon representation of the FRAP technique is shown in Fig. 2. In this manner, optical gradients may be created at specific intracellular sites or in whole cells to measure transport in a nondestructive, repetitive manner. A fuller description of FRAP instrumentation may be found in a number of technique-oriented articles (Koppel *et al.*, 1976; Koppel, 1979; Koppel *et al.*, 1980; Kapitza and Jacobson, 1986). Considering the commercial availability of hydrophobic and hydrophilic fluorescent probes, the appropriate reporter molecule can either be purchased or synthesized for transport measurements.

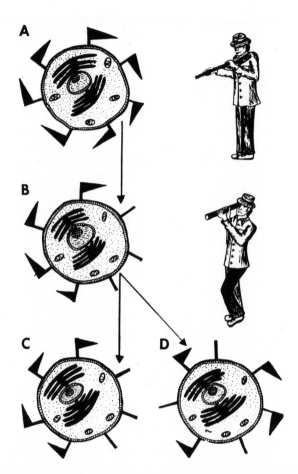

FIGURE 2. "Shooting gallery" representation of fluorescence redistribution after photobleaching. Flag structures on the surface of a cell represent a fluorescent probe bound to a membrane protein. Person with gun represents a laser beam that is utilized at high intensity to photochemically destroy the fluorescence (photobleach) in a small patch of the membrane surface (A). Following the photobleaching "shot" some flags are lost, representing the destruction of fluorescence. Person now has a spyglass rather than a gun. This represents the lower-intensity laser beam utilized to excite rather than photobleach fluorescence for the purpose of monitoring of fluorescence emission by photon detection means (B). Two outcomes of photobleaching are now possible. In one instance, the photobleached area remains nonfluorescent, suggesting no lateral mobility for probed molecules (C). The other possibility is that through the lateral mobility of probed molecules, the photobleached area can now lose photobleached molecules and capture unbleached molecules. Over a period of time, this molecular redistribution will result in a return of fluorescence to the formerly bleached nonfluorescent membrane patch (D). Following the kinetics of this redistribution leads to the determination of diffusion coefficients.

3. THE DYNAMIC PARAMETER—DETERMINATION OF ARCHITECTURE AND SIZE

Measurements of molecular transport are usually performed to explore the relationship between the structure of a transport element and its functional efficiency. The numbers obtained are related to some particular metabolic or activational activity, i.e., transmembrane glucose transport, Na/K$^+$ exchange. However, such rate measurements may also be utilized to provide a physical map of the structural impediments encountered by molecules as they move through a molecular web of interacting polymers in two or three dimensions (Schindler *et al.*, 1980; Wojcieszyn *et al.*, 1981; Luby-Phelps *et al.*, 1986), or passing through a size-dependent transport channel (Paine *et al.*, 1975; Peters, 1983; Jiang and Schindler, 1987). Molecules differing in Stokes radii or overall charge may be tracked as they diffuse through membrane channels (Paine *et al.*, 1975, Peters, 1983; Jiang and Schindler, 1987) or the cytomatrix (Wojcieszyn *et al.*, 1981; Luby-Phelps *et al.*, 1986). A relationship may then be established between the physical properties of the probe, i.e., size and charge, and the rate of movement that can be utilized for the evaluation of functional transit paths. Such measurement may be pursued under a variety of cell states to reflect subtle or gross changes in cellular architecture in a noninvasive, nondestructive manner. The measurements to be discussed utilize rate determinations, not to explore functional efficiency, but to probe geometry; in this way, time may be used to dissect space (Fig. 3).

3.1. Lateral Mobility in Biomembranes—Dynamic Analysis of Cortical Cytoskeletal Organization

Work of the past 10 years has demonstrated that plasma membrane proteins not only have physical interactions with membrane-associated cyto-

FIGURE 3. The use of time to dissect space. Measuring the rate of transport for graded size molecules through junctions, pores, transporters, or polymers can be utilized to define the diameter and geometry of functional transport channels. As represented in the top illustration, it takes longer for a defined molecule to move from one side of a channel to the other when the geometry and obstacles effectively decrease the free diffusion rate. The lower illustration demonstrates faster transit when no obstacles exist.

skeletal assemblies (Painter and Ginsberg, 1982; Landreth *et al.*, 1985; Roos *et al.*, 1985), but also may indirectly be associated with transcytoplasmic cytoskeletal networks (Berke and Fishelson, 1976; Otteskog *et al.*, 1981) (Fig. 4). Potential consequences of these interactions are: (1) protein lateral mobility and topology may be affected by internal rearrangements and modifications of the cytoskeleton (Edelman, 1976; Koppel *et al.*, 1981; Tank *et al.*, 1982), and (2) reorganization of plasma membrane receptors through such processes as cell attachment or ligand-mediated aggregation may produce changes in the assembly state and structure of cortical membrane-associated cytoskeletal assemblies (Edelman, 1976), as well as the reorganization of stress fibers, microfilament and microtubule networks (Carboni and Condeelis, 1985; Pasternak and Elson, 1985). Such a potential bidirectional signaling arrangement could link plasma membrane receptors to transcytoplasmic cytoskeletal networks through a dynamic cortical cytoskeleton. Measurements of the lateral mobility of specific membrane proteins, or a heterogeneous class of membrane glycoproteins may then serve as a nondestructive dynamic analysis of (1) the organization and dynamic rearrangements of cytoskeletons, (2) the mosaic arrangement of membrane regions

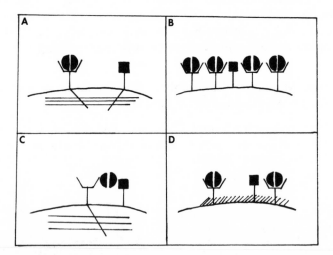

FIGURE 4. Control of plasma membrane receptor lateral mobility. A number of mechanisms have been proposed to explain the "hindered" lateral movement of transmembrane proteins in membranes. One possibility is that the cytoplasmic tail of transmembrane proteins may interact with submembranous cytoskeletal assemblies such as microtubule and microfilament networks (A). Other possibilities are that: barriers or "corrals" comprised of proteins may exist in the membrane, inhibiting diffusion (B); a receptor that is normally mobile may link to an immobile receptor (C); and/or a high surface viscosity, which may partially be a result of extracellular interactions, slows molecules down (D).

into distinct diffusion domains, and (3) the nature of protein anchoring to the membrane. Two examples of such dynamic approaches as they relate to investigations on the erythrocyte cytoskeleton and the organization of the rat liver nuclear envelope are presented below.

3.1.1. Cortical Cytoskeletal Control of Membrane Protein Lateral Mobility

To test the model that the peripheral spectrin–actin network in erythrocytes (Bennett, 1985) can impede the two-dimensional diffusion of transmembrane proteins in erythrocyte plasma membranes, two types of mouse red blood cells were employed, normal (S^+) and spectrin-deficient (S^-) spherocytic mouse cells. Each intact cell type was incubated with the fluorophore, dichlorotriazinylaminofluorescein (DTAF), at pH 10, resulting in the labeling of a single protein species, the Cl^- anion transport channel found in both cell types. When the labeled cells were scanned with a low-intensity focused laser beam (beam diameter 1 μm), a fluorescence distribution was observed as presented in Fig. 5. Following two or three such scans with a low-intensity laser beam, a higher energy ($\times 5000$) photobleaching beam was

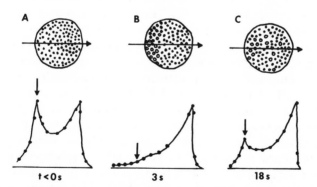

FIGURE 5. Fluorescence photobleaching the edge of mouse erythrocytes. Dots on circular projection of mouse erythrocyte plasma membrane surface represent fluorescently derivatized Cl^- anion transport channel. The arrow through the circle represents the path of a 1-μm-diameter argon laser beam ($\lambda_{ex} = 488$ nm). The double-peaked structure below the erythrocyte representation is an intensity profile of the surface fluorescence pattern. The vertical arrow over one peak is the point at which fluorescence photobleaching will occur (A). A bleach has been performed and the surface fluorescence intensity is monitored after 3 sec. The open circles in the cell representation are the bleached molecules. The edge of the fluorescence intensity profile has disappeared following the bleach pulse (B). After 18 sec, fluorescence redistribution occurs as shown by the intermixing of open circles and dots, while the bleached peak on the surface intensity profile is regaining fluorescence intensity (C).

introduced at the edge of the cell, resulting in an asymmetrical distribution of fluorescence. Subsequent scans at nonbleaching intensity show fluorescence redistribution as a function of time. Analysis of diffusion for both cell types clearly shows that the S^- mutant cell demonstrates practically unhindered membrane diffusion for the Cl^- channel ($2.5 \pm 0.6 \times 10^{-9}$ cm^2/sec), while the normal S^+ cells demonstrate a diffusion coefficient that is approximately 50 times slower ($4.5 \pm 0.8 \times 10^{-11}$ cm^1/sec) (Sheetz *et al.*, 1980; Koppel *et al.*, 1981). A further consequence of these measurements is that an effective matrix surface viscosity may be derived that is consistent with the reported viscoelastic mechanical properties of erythrocyte membranes (Koppel *et al.*, 1981). In this manner and through these types of investigations, a relationship may be established between the lateral mobility of proteins in membranes and the deformation characteristics of the membrane. A dynamic measurement leads to a better understanding of membrane mechanics and architecture.

3.1.2. Outer and Inner Nuclear Membranes—Dynamic Dissection of Structure

The cell nucleus is wrapped in a double membrane system that converges at the periphery of the morphologically distinct nuclear pore complexes (Kessel, 1973; Franke, 1974). The outer nuclear membrane faces the cytoplasm and appears to merge with endoplasmic reticulum (ER) at unique contact sites. The inner nuclear membrane, in contrast, is in close association with a dense proteinaceous lamella or nucleocytoskeleton that is predominantly comprised of a series of intermediate filament homologous polypeptides termed lamins (Gerace and Blobel, 1982; McKeon *et al.*, 1986). This lamin network forms the nuclear cage. Considering the differences in cytoskeletal attachment observed by electron microscopy between the two membranes (Kessel, 1973; Franke, 1974), a study was initiated to explore mobility differences of wheat germ agglutinin (WGA) glycoprotein receptors in the outer and inner nuclear membrane. Rat liver nuclei were isolated and incubated with fluorescein-derivatized WGA. These nuclei were subsequently washed and FRAP analysis was performed as described for spherical erythrocytes (Sheetz *et al.*, 1980; Koppel *et al.*, 1980, 1981). In addition, outer nuclear membrane-depleted nuclei were prepared and also labeled with fluorescein-derivatized WGA (Schindler *et al.*, 1985). A comparison of diffusion coefficients for whole and outer membrane-depleted nuclei suggests two dynamically different membrane systems. WGA receptors in the outer membrane were dynamic, while such receptors on the inner membrane were immobile. These observations are consistent with the possibility that the phospholipid, matrix, heterochromatin interface of the inner nuclear membrane may affect mobility in a manner analogous to submembranous membrane structures described above for erythrocytes (Sheetz *et al.*, 1980; Koppel *et al.*, 1981). In the case of the nucleus, the submembranous elements would be far more densely configured, enhancing the matrix-mediated immobilization of membrane proteins completely. The outer membrane, on the other hand, is dynamic, and may thus serve as a communication pathway between ER and nuclei.

3.2. Plant Cell Wall Porosity—Ins and Outs

The plant cell wall is an integrated meshwork of long-chain polysaccharides interspersed with glycoproteins. Functionally, it acts as a pressure-resistant container and protective barrier surrounding and in contact with the plant cell plasma membrane (Albersheim *et al.*, 1973). Macromolecules and low-molecular-weight metabolites that must interact with plasma membrane receptors or pass across the membrane, in addition to cytoplasmic molecules destined for secretion into the environment, require transport paths through the cell wall barrier. Previous studies suggest channels of sufficient size to accommodate molecules of ~60,000 Da (Tepfer and Taylor, 1981), while other investigators reported cell-wall pores to be considerably smaller (Carpita, 1982). The measurement techniques previously developed to measure pore diameter were either destructive or seriously limited in sensitivity. To overcome these deficiencies, a fluorescence-based approach was developed that permitted rapid and nondestructive analysis of transwall channel properties under a variety of growth conditions. The method is schematically represented in Fig. 6. Plant cells grown in suspension culture (Fig. 6A,B) are suspended in a high osmoticum solution containing a fluorescently derivatized (fluorescein or rhodamine) reporter macromolecule, e.g., protein, polysaccharide. The high osmoticum results in the retraction of the plant cell from the wall to form a protoplast (membrane-delimited compartment) and a large volume between the membrane and the rigid cell wall. Probe molecules capable of transwall permeation will be observed to occupy this volume between the plasma membrane and cell wall until an equilibrium is achieved (Fig. 6C). If, however, the fluorescent probe is too large for transit, reflecting a size dependency of the transport channel, a nonfluorescent cellular volume appears surrounded by a sea of fluorescence when the sample is excited by the appropriate wavelength of light (Fig. 6D). In this manner, an evaluation may be made with regard to the exclusion properties of cell wall transit paths. A typical experiment is presented in Fig. 7. To fine-tune these measurements, fluorescence photobleaching may be employed (Fig. 7). In these experiments, a focused laser beam is repetitively scanned across the volume between the cell wall and the plasmolyzed cell

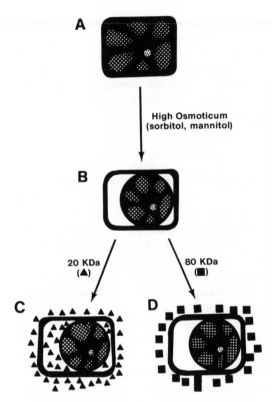

FIGURE 6. Examination of transwall transport in plant cells using fluorescent reporter molecules. Plant cells (A) are incubated in a high osmoticum containing fluorescent reporter molecules. Each sample contains a fluorescent probe of a different molecular weight. Owing to the high osmoticum, plasmolysis occurs resulting in the visualization of an intracellular volume between the plasma membrane and the cell wall (B). If the fluorescent probes are capable of permeation, as is the case for 20-kDa fluorescent dextrans, then fluorescence will be observed surrounding the cell wall and in the intracellular volume between the plasma membrane and the cell wall (C). When permeation does not occur, only extracellular localization of fluorescence is observed, as in the case of 80-kDa dextran (D).

plasma membrane [indicated as a line (scanning beam) in Fig. 7]. The fluorescent probe macromolecules, fluorescein-derivatized dextrans (F1-dextrans), which are located in this volume are photobleached by a high-intensity burst of laser light as previously described (Jiang and Schindler, 1986; Wade *et al.*, 1986; Baron-Epel *et al.*, 1988a,b). Following photobleaching, the redistribution of unbleached and bleached F1-dextrans is monitored as recovery of fluorescence. When the data are analyzed (Jiang and Schindler, 1986; Wade *et al.*, 1986; Baron-Epel *et al.*, 1988a,b), a rate constant for transwall probe transit may be obtained. As demonstrated by Peters (1983) for transnuclear transport, the data may be represented by the relationship:

$$\frac{F(-) - F(t)}{F(-) - F(0)} = a_1 e^{-k_1 t} + a_2 e^{-k_2 t} \tag{1}$$

where $F(-)$, $F(0)$, and $F(t)$ represent fluorescence signals before (prebleach), after (zero time), and at time t after photobleaching. A particularly attractive feature of this fluorescence assay is that the measurement is fast, permitting a

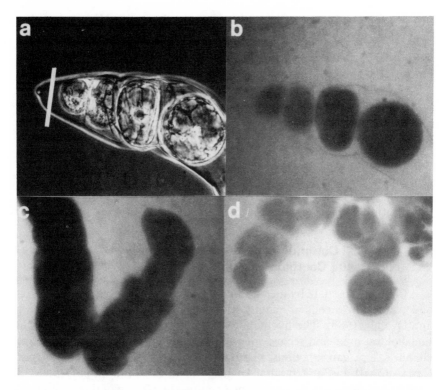

FIGURE 7. Fluorescence redistribution after photobleaching to measure transwall permeation rates. Soybean root cells (SB-1 cell line) grown in suspension culture for 3–4 days are plasmolyzed with 0.5 M mannitol. Fluorescein dextrans (10 μg/ml) were added to the culture medium at room temperature. A phase picture of plasmolyzed cells is presented in panel a, while panel b is a fluorescent view of cells following a 10-min incubation with 17.0-kDa fluorescein dextran; panel c is a fluorescent view of cells incubated with 41-kDa fluorescein dextran for 2.5 hr (×400). The line in panel a represents the path of the laser beam.

large number of measurements on the same or other cells. Such measurements can then be performed following specific cellular treatments to explore the result on cell wall porosity. In this manner, the biochemical nature of the transit channel may be explored using a dynamic approach. Results of such analysis show that major changes in cell wall porosity appear to be uniquely obtained following mild pectinase treatment. The experiments provide evidence that macromolecules with Stokes radii ≤ 3.3 nm may penetrate the cell wall unhindered. While significant hindrance is observed for macromolecules with Stokes radii of 3.3–4.6 nm, no transport is observed for macromolecules with Stokes radii > 4.6 nm. It also appears that the sieving properties of the wall, in major part, are determined by pectin rather than cellulose organization (Baron-Epel *et al.,* 1988b).

Extensions of FRAP Porosity Measurements in Biological Systems

Prokaryotes and yeast are surrounded by cell wall structures whose properties may vary with growth conditions. Using these organisms and the ability for genetic manipulation, it should be possible to dynamically dissect the macromolecular structures controlling cell wall porosity. In addition, cell wall transit variants may be selected for cloning using the transport of fluorescent probe molecules as selection criteria. In this manner, a number of cells could be isolated with defects that lead to modified cell wall growth and organization.

3.3. Cell–Cell Communication—Dynamic Linkages and Tissue Coordination

Cellular cross talk by means of metabolic and ionic gradients organizes and synchronizes cellular activity in tissues (Loewenstein, 1979; Gunning and Overall, 1983). The fluctuations in transport of low-molecular-weight metabolites, ions, and second messengers through intercellular connecting channels, e.g., gap junctions, plasmodesmata, may serve as a binary code (open–closed) and/or temporal language to initiate and amplify periodic chemical events. In this manner, the opening and closing of transport channels may serve to initiate biological rhythms in tissues. These rhythms are integrated to produce larger cycles of cellular activity and organismal response. An outcome of such a view is that cellular activity in a tissue may be a function of location and surrounding connections. This may best be represented as viewed in Fig. 8. In the examples presented, two coupled cells are

A

B

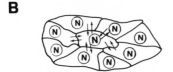

FIGURE 8. Cell–cell communication as a function of cellular position and contacts. Intercellular communication between two cells is demonstrated in A; multiple communication linkages are established for the central cells in B. N, nucleus.

observed to equally "taste" each other's cytoplasm, and the rate of information transferred between them is equal (Fig. 8A). On the other hand, as more cells are connected to form tissues, a single cell surrounded by other contiguous cells may now have a more rapid exchange of molecular information between cytoplasms because of numerous dynamic linkages (Fig. 8B). This may be the basis for more complex differences in the rate of molecular flow between populations of different biochemically homogeneous cells within the organization of the tissue. A variation in overall cellular activity would therefore be predicted to exist for cells surrounded by five, three, or two contacting cells.

To examine such dynamic linkages, approaches must be designed to make transport measurements, in a noninvasive manner, on large populations of contacting cells. Using limited modifications of the approaches discussed for trans-cell wall measurements (Section 3.2), fluorescence photobleaching may be utilized to examine the consequences of cellular position and extent of cell–cell contact for both plant and animal cell differentiation and activity. To pursue these problems, a low-molecular-weight fluorescent probe is utilized that can permeate the cell cytoplasm and traverse the connecting protein tunnels between contiguous cells. An excellent molecule for this approach has been carboxyfluorescein diacetate (Wade *et al.,* 1986; Baron-Epel *et al.,* 1988a). The significant advantage of this compound is that it can passively diffuse across cellular membranes. Upon reaching the cytoplasm, intracellular esterases release the ester-linked acetates, exposing sufficient negative charge to trap the carboxyfluorescein in the cytoplasm (Fig. 9). In this manner, all cells are uniformly labeled and available for continuous measurements (Fig. 10A). Cells are then selected for photobleaching, resulting in photochemical destruction of dye fluorescence (Fig. 10B). Redistribution of fluorescence from surrounding cells may then be monitored to examine rates and patterns of photobleaching. Mixed cellular populations may also be employed to determine the types of heterotypic cell couplings that may occur. A typical analysis demonstrates that the rate of fluorescence filling into a bleached cell is equivalent to the rate of loss of fluorescence from a contacting nonphotobleached cell (Fig. 11). Such curves may now be prepared for cell couplets, triplets, quadruplets, etc., to correlate intercellular transit rates to the number of contacting cells.

Simple to Complex Intercellular Communication—The Possibilities

Such model system approaches may lend themselves to higher-order scaling in computers to monitor the extremely complex communication patterns between heart cells that lead, under the best circumstances, to organized rhythmic activity and, in the worst cases, to the chaotic mix of disor-

A

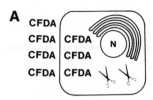

CFDA
CFDA | CFDA
CFDA | CFDA
CFDA | CFDA

B

C

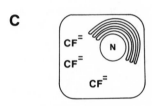

FIGURE 9. Carboxyfluorescein diacetate (CFDA) as a permeant probe for cell–cell communication. Diacetylated carboxyfluorescein is capable of transmembrane permeation gaining access to the cytoplasm (A). Cytoplasmic esterases, schematically represented as scissors, hydrolyze the ester-linked acetates (B), creating negatively charged carboxyfluorescein which is trapped because of the newly acquired charge in the cytoplasm (C).

A

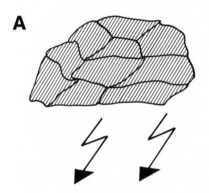

B

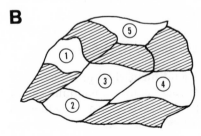

FIGURE 10. Fluorescence photobleaching of carboxyfluorescein-labeled cells in tissue culture. Cells grown in tissue culture are uniformly cytoplasmically labeled with carboxyfluorescein as described in Fig. 9. Particular cells may then be chosen for photobleaching and monitored to obtain transport flux rates. Multiple measurements may be concurrently monitored, and experiments may be repeated on the same cells.

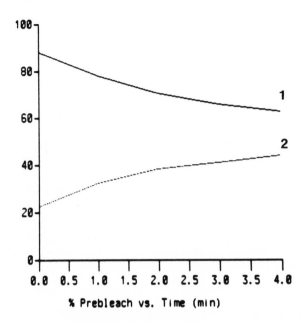

FIGURE 11. Intercellular communication between two contacting cells. As would be expected by the law of mass action, the rate of gain in fluorescence of a photobleached cell should be reflected by the rate of loss in fluorescence of the nonphotobleached contacting cell. This is observed by the fluorescence redistribution curves showing that the rate of loss of fluorescence intensity (1) is approximately equal to the rate of gain (2). The data are obtained using the cell–cell communication algorithm of the Meridian Instruments, Inc. ACAS 470 Interactive Laser Cytometer.

ganized arrhythmic pumping that leads to death. Another important advantage of these fluorescence real-time approaches is that communication patterns may be viewed over the long term in developmental systems such as embryos. The changing patterns of cellular communication may be probed by photobleaching, starting from the two-cell stage to multiple cells. Such approaches may provide clues to the nature of differentiating substances (morphogens) and to the complex gradients that have been suggested to trigger steps in embryonic development and differentiation.

3.4. Transnuclear Transport—Dynamic Analysis of the Nuclear Pore Complex

The chromatin is isolated from the cytoplasmic compartment by the intervention of a proteinaceous structure termed the nuclear lamina (Section 3.1.2). This barrier is not complete, but contains transit channels of specific

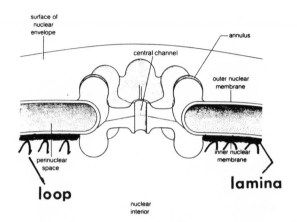

FIGURE 12. Schematic representation of the nuclear pore complex in eukaryotic cells. Structures shown as loops are DNA attachment sites to the peripheral nuclear karyoskeleton (lamina). The central channel is speculated to be the bidirectional transnuclear transport channel.

geometry (Fig. 12) that permit transnuclear migration of mRNA into the cytoplasm for protein synthesis and also protein transport into the nucleus for nuclear activation or structural organization and assembly. Questions relating to channel activation and structure of the nuclear pore complex are difficult to treat by traditional biochemical approaches because of problems relating to nuclear pore complex isolation, purification, and reconstitution. In an attempt to circumvent a number of these problems, we have attempted to tackle structure using dynamic measurements, both *in vitro* and *in vivo*. Considering that the nuclear pore complex is the only nuclear structure permitting transnuclear transport, measurements can be made using molecules of defined charge and geometry to ascertain variations in transit rate. Such measurements, in conjunction with classical diffusion equations, then permit an estimation of the functional channel diameter (Paine *et al.,* 1975; Peters, 1983; Jiang and Schindler, 1986). In a similar manner, antibodies that are prepared against nuclear pore complex polypeptides (Schindler and Jiang, 1986) may be utilized to explore their effects on macromolecular transport. Such information may be used to define polypeptide placement or, potentially, enzymatic activity within the nuclear pore complex.

Nuclear FRAP

The technique to be discussed for measuring transnuclear transport has been employed for both *in vitro* measurements on isolated rat liver nuclei and *in vivo* measurements on Balb/c 3T3 fibroblasts. For *in vitro* measurements, the fluorescent dextran influx assay was performed in the following manner.

Nuclei were suspended in 1 ml of 0.25 M sucrose/10 mM Hepes/1 mM Mg^{2+} (pH 7.4) buffer that contained 1-μM fluorescein-labeled 64-kDa dextrans (Mw/Mn < 1.25). The nuclei were incubated and equilibrated in the dextran solution for 30 min at room temperature. A 5-μl aliquot of this nuclear suspension was placed on a slide, and a coverslip was placed on top of the sample. Melted paraffin was used to seal the coverslip to the slide. Photobleaching experiments were essentially done as described elsewhere (Peters, 1983; Jiang and Schindler, 1986, 1987). Experiments by Peters (1983) demonstrated that neither the size (although it should have a diameter greater than that of the nucleus) nor the intensity profile of the laser beam is critical. The beam in the focal plane covers a major part of the cross section of the nucleus, which ensures that a photobleach will maximally deplete the fluorescence in the nucleus. After a nuclear photobleach, redistribution of fluorescence between the nuclear compartment and solution was calculated by a method of Peters (1983). Considering that diffusion across the membrane is the rate-limiting step for dextran influx, transport kinetics then follow the equation

$$\frac{Ce - C(t)}{Ce - C(0)} = e^{-kt} \tag{2}$$

where Ce, $C(0)$, and $C(t)$ are tracer concentrations in the nucleus at equilibrium, zero time, and time t, respectively. Rate constant k is related to permeability coefficient P by the equation $P = (V/A)k$, in which V and A are the volume and area, respectively, of the nucleus. As with all photobleaching experiments, fluorescence intensities measured are representative of the fluorescence-derivatized solute concentration at any particular time (t). Accordingly, Eq. (2) may be represented as

$$\frac{F(-) - F(t)}{F(-) - F(0)} = e^{-kt} \tag{3}$$

where $F(-)$, $F(0)$, and $F(t)$ are the fluorescence signals before (prebleach), after, and at time t after photobleaching. As demonstrated by Peters (1983), the data are satisfactorily plotted as

$$\frac{F(-) - F(t)}{F(-) - F(0)} = a_1 e^{-k_1 t} + a_2 e^{-k_2 t} \tag{1}$$

or the sum of two exponentials. In all instances, multiple bleaches demonstrated the same recovery profiles, which suggests no major photochemical damage. Prebleach emission pattern of a nucleus in equilibrium with a solution of fluorescently labeled dextran (64 kDa) is shown in Fig. 13A. The ordinate is an arbitrary scale of emitted fluorescence intensity, whereas the abscissa is the scan point across a cell. A scan period is the time required by the laser beam to traverse the sample field (5–10 sec). After the prebleach scan, the laser intensity is increased by ~5000, which results in the photo-

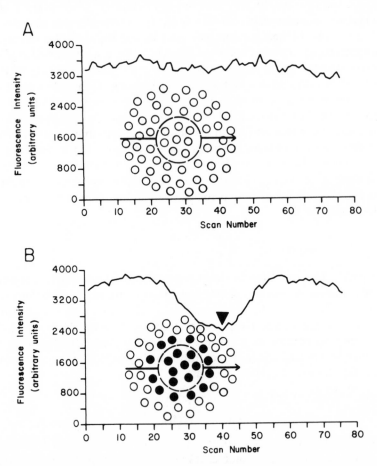

FIGURE 13. Fluorescent dextran transport as measured by FRAP. (A) Laser beam-excited prebleach emission pattern of a nucleus in equilibrium with a solution of fluorescently labeled dextran (64 kDa). The ordinate is an arbitrary scale of emitted fluorescence intensity; the abscissa represents the scan number. A scan is the time required by the laser beam to traverse the sample field (5–10 sec). After a prebleach scan, the laser intensity is increased by ~5000, resulting in the photobleaching of the entire nucleus. This is observed as an intensity dip in B. Recovery of the dip to prebleach levels is proportional to the dextran flux (Peters, 1983). The diagrams in A and B represent the bleaching process. The open circles in A are exaggerated fluorescent dextran molecules within and surrounding the nucleus. The arrow represents the laser beam scan path. The solid circles in B represent the bleached dextrans.

bleaching of the entire nucleus. This is represented by the intensity dip observed in Fig. 13B (arrowhead). The recovery of this dip to prebleach levels is proportional to the dextran flux rate (Peters, 1983; Jiang and Schindler, 1986). The diagrams in A and B (Fig. 13) represent the bleaching process. The open circles in A represent fluorescent dextran molecules within and surrounding the nucleus. The line represents the path of the expanded 6-μm beam. The solid circles in B represent the bleached dextrans.

The rate constants (k_1 and k_2) obtained by two-component linear fitting may be further employed to calculate either the number of transport channels or the radius of the channel using the equation:

$$P = nDA/X \qquad (4)$$

where n represents the area density of channels, D is the diffusion coefficient, A is the effective pore or transport channel area accessible to the solute, and X is the channel length. In this manner, rates may again be utilized to pursue structural information on transit channel properties in a nondestructive manner.

4. CONCLUSIONS

In this chapter, we have attempted to demonstrate how the measurement of molecular movement on the cell surface and in the cytoplasm may be utilized to provide information on the structure of cellular compartments and components. The value of these approaches using fluorescence photobleaching measurements and fluorescent analogues of metabolites, signaling molecules, and structural elements is that nondestructive measurements may be made at the level of the single living cell or tissue. The rapid development and commercial availability of fluorescent probes and analytical instrumentation suggest that the dynamic parameter will become a generally available and valuable new tool for all laboratories pursuing problems in cell biology.

ACKNOWLEDGMENTS. We thank Dr. Margaret H. Wade (Meridian Instruments, Inc., Okemos, Mich.) for helpful discussions and Sally Meiners for help with the artwork. The work reported was funded in part by NIH Grant GM30157. P.K.G. thanks the Nitrogen Availability program at Michigan State University for fellowship support.

5. REFERENCES

Albersheim, P., Bauer, W. D., Keegstra, K., and Talmadge, K. W., 1973, in *Biogenesis of Plant Cell Wall Polysaccharides* (F. Loewus, ed.), Academic Press, New York, pp. 117–147.

Baron-Epel, O., Gharyal, P. K., and Schindler, M., 1988a, *Planta,* **175:**389–395.

Baron-Epel, O., Hernandez, D., Jiang, L.-W., Meiners, S., and Schindler, M., 1988b, *J. Cell Biol.* **106:**715–721.

Bean, B. P., Nowycky, M. C., and Tsien, R. W., 1983, *Nature* **30:**371–375.

Bennett, V., 1985, *Annu. Rev. Biochem.* **54:**273–304.

Berke, G., and Fishelson, Z., 1976, *Proc. Natl. Acad. Sci. USA* **73:**4580–4583.

Carboni, J. M., and Condeelis, J. S., 1985, *J. Cell Biol.* **100:**1884–1893.

Carpita, N., 1982, *Science* **218:**813–814.

Conti-Tronconi, B. M., and Raftery, M. A., 1982, *Annu. Rev. Biochem.* **51:**491–530.

Edelman, G. M., 1976, *Science* **192:**218–226.

Franke, W. W., 1974, *Philos. Trans. R. Soc. Lond. Ser. B* **268:**67–93.

Gerace, L., and Blobel, G., 1982, *Cold Spring Harbor Symp. Quant. Biol.* **46:**967–978.

Gunning, B. E. S., and Overall, R. L., 1983, *Bioscience* **33:**260–265.

Hondeghem, L. M., and Katzung, B. G., 1984, *Annu. Rev. Pharmacol. Toxicol.* **24:**387–423.

Jiang, L.-W., and Schindler, M., 1986, *J. Cell Biol.* **102:**853–858.

Jiang, L.-W., and Schindler, M., 1987, *Biochemistry* **26:**1546–1551.

Kapitza, H.-G., and Jacobson, K. A., 1986, in *Techniques for the Analysis of Membrane Proteins* (C. I. Ragan and R. J. Cherry, eds.), Chapman & Hall, London, pp. 345–375.

Kessel, R. G., 1973, *Prog. Surf. Membr. Sci.* **6:**243–329.

Koppel, D. E., 1979, *Biophys. J.* **28:**281–292.

Koppel, D. E., Axelrod, D., Schlessinger, J., Elson, E. L., and Webb, W. W., 1976, *Biophys. J.* **16:**1315–1329.

Koppel, D. E., Sheetz, M. P., and Schindler, M., 1980, *Biophys. J.* **30:**187–192.

Koppel, D. E., Sheetz, M. P., and Schindler, M., 1981, *Proc. Natl. Acad. Sci. USA* **78:**3576–3580.

Koshland, D. E., Jr., 1981, *Annu. Rev. Biochem.* **50:**765–782.

Landreth, G. E., Williams, L. K., and Rieser, G. D., 1985, *J. Cell Biol.* **101:**1341–1350.

Loewenstein, W. R., 1979, *Biochim. Biophys. Acta* **560:**1–65.

Luby-Phelps, K., Taylor, D. L., and Lanni, F., 1986, *J. Cell Biol.* **102:**2015–2022.

McKeon, F. D., Kirschner, M. W., and Caput, D., 1986, *Nature* **319:**463–468.

Metcalf, T. N., III, Wang, J. L., Schubert, K. R., and Schindler, M., 1983, *Biochemistry* **22:**3969–3975.

Mitchell, P., 1979, *Euro. J. Biochem.* **95:**1–20.

Nikaido, H., and Nakae, T., 1979, *Adv. Microb. Physiol.* **20:**163–250.

Noma, A., 1983, *Nature* **305:**147–148.

Otteskog, P., Ege, T., and Sundquist, K.-G., 1981, *Exp. Cell Res.* **136:**203–213.

Paine, P. L., Moore, L. C., and Horowitz, S. B., 1975, *Nature* **254:**109–114.

Painter, R. G., and Ginsberg, M., 1982, *J. Cell Biol.* **92:**565–573.

Pasternak, C., and Elson, E., 1985, *J. Cell Biol.* **100:**860–872.

Peters, R., 1983, *J. Biol. Chem.* **258:**11427–11429.

Reuter, H., Cachelin, A. B., dePeyer, J. E., and Kokubun, S., 1983, *Cold Spring Harbor Symp. Quant. Biol.* **48:**193–200.

Roos, E., Spiele, H., Feltkamp, C. A., Huisman, H., Wiegart, F. A. C., Traas, J., and Meland, D. A. M., 1985, *J. Cell Biol.* **101:**1817–1825.

Schindler, M., and Jiang, L.-W., 1986, *J. Cell Biol.* **102:**859–862.

Schindler, M., Osborn, M. J., and Koppel, D. E., 1980, *Nature* **283:**346–350.

Schindler, M., Holland, J. F., and Hogan, M., 1985, *J. Cell Biol.* **100:**1408–1414.

Schindler, M., Trosko, J. E., and Wade, M. H., 1987, *Methods Enzymol.* **141:**439–447.

Schindler, M., Jiang, L.-W., Swaisgood, M., and Wade, M. H., 1990, *Methods Cell Biol.* **32:**423–445.

Sheetz, M. P., Schindler, M., and Koppel, D. E., 1980, *Nature* **285**:510–512.

Tank, D. W., Wu, E.-S., and Webb, W. W., 1982, *J. Cell Biol.* **92**:218–226.

Tepfer, M., and Taylor, I. E. P., 1981, *Science* **213**:761–763.

Wade, M. H., Trosko, J. E., and Schindler, M., 1986, *Science* **23**:525–528.

Wojcieszyn, J. W., Schlegel, R. A., Wu, E.-S., and Jacobson, K. A., 1981, *Proc. Natl. Acad. Sci. USA* **78**:4407–4410.

Index

283